VISIT SUNNY CHERNOBYL

Andrew Blackwell has worked as film editor, story consultant, and reporter on projects featured on NPR, PBS, the BBC, the *New York Times* online, and in the weekly news magazine *Dan Rather Reports*. He lives in New York City.

Praise for *Visit Sunny Chernobyl*

'This book is based on a simple but brilliant idea: to hang out in the most polluted and toxic places on the planet. Well it's a brilliant idea for a book at any rate, if not a holiday'
Mail on Sunday

'[Andrew Blackwell] has seen the worst we can do to ourselves, met the best of us trying to repair the damage, and discovered – mystifyingly at times – beauty in the planetary dark side'
The Times

'This is one of the best concepts for a non-fiction book I have come across in some time ... There is a great deal of acute observation, proper thinking and challenging material in each of the journeys' *Scotland on Sunday*

'An essential read. A very funny – and very disturbing – look at some parts of our world that need to be acknowledged. *Visit Sunny Chernobyl* is my new favorite guidebook to some places I admit to have visited'
Peter Greenburg, Travel Editor, CBS News

'A certain amount of depression and anger does result from reading Andrew Blackwell's *Visit Sunny Chernobyl*, but Blackwell is such a good and amusing travel writer, such an engaging companion arou   come away
son

D0306711

'Blackwell is a very good writer, with laconic, graphic and gently ironic style, at times reminiscent of both Scott Fitzgerald and Ernest Hemingway . . . *Visit Sunny Chernobyl* inspires rather than scares' *Engineering & Technology Magazine*

'A wise, witty travel adventure that packs a punch - and one of the most entertaining and informative books I've read in years. *Visit Sunny Chernobyl* is a joy to read and will make you think' Dan Rather

'Every now and again a ray of sunshine lifts the usual gloom-and-doom of environmental crises and this witty, warm and refreshingly honest tour of the netherworld of modern life offers a particularly bright one . . . It is a moving and often hilarious story of human dignity rising above unimaginable squalor' David Shukman, Science Editor, BBC News, and author of *An Iceberg as Big as Manhattan*

VISIT SUNNY CHERNOBYL

ADVENTURES IN THE WORLD'S MOST

POLLUTED PLACES

Andrew Blackwell

arrow books

London Borough of Southwark	
J	
SK 2341842 7	
Askews & Holts	14-May-2013
363.731 NAT	£7.99

Andrew ⋯⋯⋯⋯⋯⋯⋯⋯⋯ ⋯ight under the Copyrigh⋯ Designs
and P⋯⋯⋯⋯⋯⋯⋯⋯⋯⋯⋯⋯⋯ ⋯ntified as the author of ⋯⋯is work

This book is sold subject to the condition that it shall not, by way of trade
or otherwise, be lent, resold, hired out, or otherwise circulated without the
publisher's prior consent in any form of binding or cover other than that in
which it is published and without a similar condition, including this condition,
being imposed on the subsequent purchaser.

First published in the United States in 2012 by Rodale Inc., New York

First published in Great Britain in 2012 by Random House Books
Random House, 20 Vauxhall Bridge Road,
London SW1V 2SA

www.randomhouse.co.uk

Addresses for companies within The Random House Group Limited can be
found at: www.randomhouse.co.uk/offices.htm

The Random House Group Limited Reg. No. 954009

A CIP catalogue record for this book
is available from the British Library

ISBN 9780099549642

The Random House Group Limited supports the Forest Stewardship
Council® (FSC®), the leading international forest-certification
organisation. Our books carrying the FSC label are printed on
FSC®-certified paper. FSC is the only forest-certification scheme
supported by the leading environmental organisations, including
Greenpeace. Our paper procurement policy can be found at:
www.randomhouse.co.uk/environment

Printed and bound by CPI Group (UK) Ltd, Croydon, CR0 4YY

CONTENTS

AUTHOR'S NOTE

This is a work of nonfiction. I have changed names when it seemed appropriate and made occasional, immaterial rearrangements of chronology. While allowing minor editing for clarity, I have striven to put words in quotes only when they were actually said—in that form, by that person. I have tried always to leave outside of quotation marks anything that I chose to rephrase, whether because of the limitations of my notes and memory, or for purposes of brevity, or due to the grey areas of on-the-fly translation. When so imprudent as to include facts or figures, I have attempted to be scrupulous in my choice and interpretation of sources.

For photographs, further maps, and more information about the places, people, organizations, and issues touched on in this book—or if you think I got anything wrong and would like to tell me so—please visit:

www.visitsunnychernobyl.com

"EVEN WHAT IS MOST UNNATURAL IS PART OF NATURE."

—GEORG CHRISTOPH TOBLER, "DIE NATUR"

PROLOGUE

We come in smooth, coasting, sliding between stands of reeds, water lapping against the metal sides of the rowboat. A host of dragonflies dances around us. They land on the dented edges of the boat, on the oar handles, on my hands. A little one lands on my nose. As I row, the oars catch on thick colonies of lily pads, rising out of the water ripe and green. They glisten for a moment in the midday sun, then plunge back under with the following stroke.

Sitting on the bow, a young woman gazes out over the water, over an expanse of marshy islands extending in every direction. The air vibrates with a shrill chorus of frogs.

A high beep breaks the reverie. Olena looks down at the device in her hand. It's a radiation detector. We bought it in Kiev a few days ago.

"It says twenty," she calls out. "I guess we're going in the right direction."

I look down at the tattered photocopy of a map, handed to me onshore by a smiling weekend fisherman. It shows part of the Kiev Sea, a broad, placid reservoir that greets the confluence of the Pripyat and Dnieper Rivers. I'm almost certain we've already reached our destination. Here, somewhere among the dragonflies and lily pads, we have crossed a boundary. A border guarded only by coasting herons and warbling frogs.

We have just infiltrated the world's most radioactive ecosystem.

This is the Exclusion Zone, site of the infamous Chernobyl disaster. A radiological quarantine covering more than a thousand square miles of Ukraine and Belarus, it is largely closed to human activity, even a quarter century after the meltdown. Entry to the zone is forbidden without prior permission, an official escort, and a sheaf of paperwork. A double fence of concrete posts and barbed wire encircles it, and guards man the entrances.

On land, that is. By water, the zone is open for your enjoyment. Or if not exactly *open*, not so closed that anything stands in the way of an afternoon paddle. All you need is a rowboat and some way to get to Strakholissya, the town closest to where the Pripyat River flows out of the zone. You might not even need your own rowboat. In Strakholissya, after a picnic of strawberries and sandwiches, we met an old woman who let us make off with hers. When we asked her what it would cost, she seemed to take it as a philosophical question. "What would such a thing cost?" she said, refusing the money. We paid her in strawberries.

I keep rowing. The tall grass slides along the side. I warm myself in the sun. This is the rest of the infiltration plan. Only this. To peer at the dragonfly on my nose. To watch a vast thunderhead erupt from beyond the horizon, billowing into the sky.

\\\\\\\\\\

Years ago, I spent six months in India. I saw any number of exotic sights there, from traditional villages in a remote corner of Rajasthan to the gilt sanctuary of a Buddhist monastery perched like a citadel on the slopes of the Himalayas. I watched as traditional fishermen pulled nets full of writhing fish out of the Arabian Sea for sale right on the beach. I contemplated sacred carvings in thousand-year-old Jain temples.

You know. The usual crap.

And then there was Kanpur. Newly awarded the title of India's Most Polluted City by the national government—which says something in India—

it was not an obvious destination. In fact, hardly anybody outside India has ever heard of Kanpur, and few people inside the country would give it a second thought. But I was traveling with an environmentalist, and environmentalists can have unusual sightseeing priorities.

What followed was an intensive, three-day tour of dysfunctional sewage-treatment plants, illegal industrial dumps, poisonous tanneries, and feces-strewn beaches. The crowning moment was our visit to a traditional Hindu bathing festival in which scores of pilgrims dunked themselves in a rank stretch of the sacred—but horribly contaminated—river Ganges, collecting bottles of holy, chromium-laced water for use back home. All this, and not another tourist in sight.

Inexplicably, Kanpur became the highlight of my entire time in India. *Kanpur.* I couldn't account for it. Did I have a thing for industrial waste? Was I just some kind of environmental rubbernecker?

That wasn't it. In Kanpur, I had found something. Something I hadn't encountered anywhere else. I couldn't shake it: the sense of having stumbled into a wholly unexpected place. Of having seen something there, among the effluent pipes and the open latrines. A trace of the future, and of the present. And of something else—something inscrutably, mystifyingly beautiful.

Leaving the country, I looked up Kanpur in my guidebooks to get their take on the place. Nothing. As far as Lonely Planet and the Rough Guide were concerned, Kanpur—a city of millions—was literally not on the map. It was a noxious backwater to be avoided and ignored.

And for any traveler curious enough to find a place like Kanpur interesting, well . . . there was just no way to find out about it.

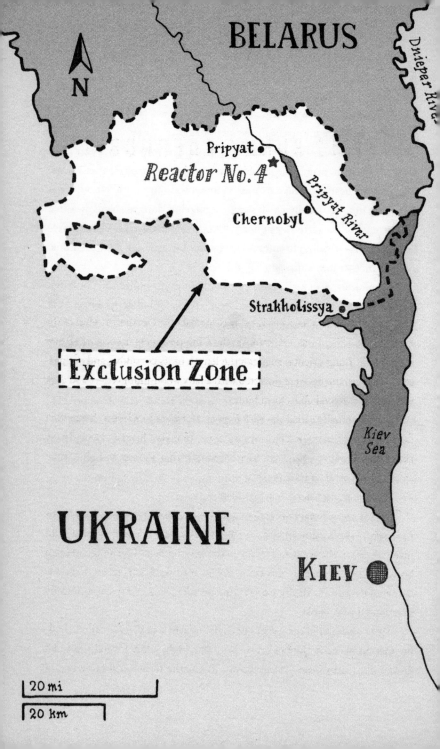

ONE

VISIT SUNNY CHERNOBYL

It began on a train. Vienna to Kiev, rocking back and forth in a cabin of the *Kiev Express*. There was a certain Agatha Christie–meets–Leonid Brezhnev charm to it. Long oriental rugs ran the length of its corridors, and the passenger compartments were outfitted with a faux wood-grain veneer and dark red seats that folded up to form bunks.

It's not actually called the *Kiev Express*. If it were an express, it wouldn't take thirty-six hours. In fact, train is no way to make this trip. I bought my ticket only because I believed, unaccountably, that Vienna and Kiev were close to each other. They are not.

I was going to Chernobyl, on vacation.

Trains are for reading, and I had brought a pair of books: *Voices from Chernobyl*, a collection of survivor interviews, and *Wormwood Forest*, an investigation of the accident's effect on the environment. I recommend them both, although when I say that trains are for reading, I don't mean that I was doing all that much. Really I was taking an epic series of naps, sporadically interrupted with books.

My companion in the passenger compartment was Max, a rotund, smiling man in his early thirties. Max spoke in a high, oddly formal voice and looked like a grown-up Charlie Brown, if Charlie Brown had grown up in the USSR. Originally from Kiev, he now worked in Australia as a computer programmer. He had an endearing way of stating the obvious. I would wake up from a nap, my book sliding onto the floor, and look out the window to see that we had stopped in a station.

"We have stopped," Max would say.

We spent the first night crossing the length of Slovakia. A beautiful dusk settled over the cracked smokestacks of deserted factories.

In the morning, we reached the Ukrainian border and rolled into a cluttered rail yard, coming to rest between a set of oversize jacks, taller than the train car itself. A team of crusty rail workers set themselves wrenching and hammering at the wheels of the train, and soon the jacks were raising the entire car into the air, leaving the wheel trucks beneath us on the rails.

The train tracks in the former Soviet Union don't match those in Europe, you see. So they were changing the wheels on the train.

"They are changing the wheels on the train," Max said.

By afternoon we had entered the flowered alpine landscape of the Carpathian Mountains, and Max had become curious about my plans. I chose not to tell him that I was embarking on an epic, years-long quest to visit the world's most polluted places. I just said I was headed for Chernobyl.

His face lit up. He had stories to tell. In the spring of 1986, when word of the disaster got out, he was eleven years old, living in Kiev. Soon, people were trying to get their children out of the city. It was nearly impossible to get train tickets, Max said, but somehow his family got him onto a train bound southeast for the Crimea. Even though tickets were so hard to come by, the train was nearly empty, and Max implied that the government had manufactured the ticket shortage to keep people from leaving the city.

"When we arrived," he said, "the train was surrounded by soldiers. They tested everyone and their things for radiation before allowing them to move on. They were trying to keep people from spreading contamination."

He stayed away from Kiev that entire summer. From his parents, he heard stories about life in the city during those months. The streets were washed down every day. Bakeries that had once left their wares out in the open on shelves now wrapped them in plastic.

Max talked about the possibility that cancer rates in the area had increased because of Chernobyl, and he told me that his wife, also from Kiev, had abnormalities in her thyroid, which he attributed to radioactive exposure.

"It's very lucky Kiev didn't get more radiation, thanks to the winds," he said. Then, in his very polite, clipped voice, he asked, "And what do you think about nuclear energy?"

That night I lay restless in my bunk and imagined—as only an American can—the post-Soviet gloom slipping by outside, felt the train shudder as it pushed through the thick ether left behind by an empire. In the book of Chernobyl survivors' stories, I read an account by a firefighter's widow. They were newly married when her husband responded to the fire at the reactor. One of the first at the scene, he received catastrophic doses of radiation and died after two weeks of gruesome illness.

Desperately in love, his wife had snuck into the hospital to accompany him in his ordeal, even though his very body was dangerously radioactive.

"I don't know what I should talk about," she says in her account. "About death or about love? Or are they the same?"

\\\\\\\\\\\\

Kiev is a beautiful city, a true Paris of the East, a charming metropolis whose forests of horse chestnut trees set off its ancient churches and classic apartment buildings like jewels on a bed of crumpled green velvet. The trick is to come in the summertime, when a warm breeze blows across the Dnieper River and the bars and cafés spill out into the gentle evening. You can stroll down the Andriyivskyy Descent, lined with cafés and shops, or explore the mysterious catacombs of the Pechersk Lavra, with its menagerie of dead monks. Or you can dive into the city's pulsing downtown nightlife.

I went straight for the Chernobyl Museum.

There's a special blend of horror and civic pride on display at any museum dedicated to a local industrial disaster, and the Chernobyl Museum is surely the best of its kind. The place incorporates history, memorial, commentary, art, religion, and even fashion under a curatorial ethos that is the mutant offspring of several different aesthetics.

In one of the museum's two main halls, I found a bizarre temple-like space. Soothing Russian choral music emanated from the walls. In the center of the room lay a full-size replica of the top face of the infamous reactor. A dugout canoe was suspended above it, heaped with a bewildering mixture of religious images and children's stuffed toys. I tried to understand the room's message, and could not. Empty contamination suits lingered in the shadows, arranged in postures of bafflement and ennui.

The second hall housed a definitive collection of Chernobyl memorabilia, as well as a tall aluminum scaffold hung with mannequins wearing nuclear cleanup gear. They seemed to be flying in formation, a squad of unusual superheroes. Their leader, arms upraised, wore a black firefighting suit with large white stripes and a metal backpack connected to a gas mask. Through the bubble of the helmet's face guard, I could just make out the cool, retail gaze of a female head, with full eyelashes and painted plastic lips.

Underneath, there was a cross-sectioned model of the reactor building in its pre-accident state. As I peered into it to get a view of the reactor's inner workings, two docents lurking by the door noticed my interest. Moving with the curt authority of guards, they rushed forward to turn the model on, groping at a control panel attached to the base. The model reactor glowed warmly, showing the normal circulation of water in the core. But the women were unsatisfied. Fussing in Ukrainian, they began flipping the switch back and forth, wiggling and slapping the little control panel with increasing fervor. Finally, they jiggled the switch just right, and the rest of the reactor's systems—water and steam pipes, cooling systems and boilers—flickered to life.

\\\\\\\\\\\

To understand the Chernobyl accident, it helps to know something about how electricity gets generated and, specifically, about nuclear power—though not so much that your eyes glaze over.

In general, power plants generate electricity by spinning turbines. Picture a big hamster wheel and you get the idea. Each turbine is connected to a generator, in which a conductor turns through the field of a strong magnet, thus creating electricity by magic. Men in hard hats then distribute this power to entire continents full of televisions and toaster ovens.

The ageless question, then, is just how to spin all those damn turbines. You can build a dam to collect huge volumes of water that you can let rush through your turbines. You can build windmills with little generators that get powered by the turning rotors. Or you can boil a lot of water and force the steam into the turbine under high pressure.

This last one works great, but you need a hell of a lot of heat to make enough steam. Where are you going to get it? Well, you can burn coal, natural gas, or even trash, if you like. That, or you can cook up some nuclear fission.

Oh, *fission*. People make it sound so complicated, but any chump can get the basics. It involves—to skip most of the physics—piling up a giant stack of purified uranium to make your reactor's core. You'll have to mix some graphite in with the uranium, to mellow out the neutrons it's emitting.

We good? Okay. Once you've got the core together, install some plumbing in it so you can run water through to carry off the heat, and then just stand back and cross your fingers.

A few of the uranium atoms in your core will spontaneously split—they're funny that way—and when they do, they'll give off heat and some neutrons. It doesn't matter if you don't know what neutrons are, other than that they're tiny and will shoot off like bullets, colliding with neighboring uranium atoms and causing *them* to split. This will give off more heat and more neutrons, which will cause still further atoms to split, and so on, and so on, and so on. The immense heat created by this chain reaction will heat the water, which will create the steam, which will spin the turbines at terrifying speed, which will turn the generators, which will create an ungodly amount of electricity, which will be used to keep office buildings uncomfortably cold in the middle of summer.

So far, so good.

The problem with this chain reaction is that, by its very nature, it tends to run out of control. So to keep your reactor's apocalyptic side in check, you should slide some rods made of boron or hafnium into the reactor core. (Remember to make room for them while you're stacking the uranium.) These rods—let's call them *control rods*—will be like sponges, absorbing all those lively, bullet-like neutrons. With the control rods duly inserted, you'll get . . . nothing.

The trick, then, is to find the happy medium, while remaining on the correct side of the line that separates air-conditioning from catastrophe. To do this, you'll need to pull the control rods out of the core far enough to let the chain reaction begin, but not so far that it runs out of control. Then you can heat water and spin turbines and generate electricity to your heart's content.

But pull the control rods out slowly, okay? And for the love of God, please—*please*—put them back when you're done.

\\\\\\\\\\\

With the Chernobyl Museum taken care of, I had a couple of days to kill in Kiev before my excursion to Chernobyl itself, and I spent them exploring my new neighborhood. I was living in style, sidestepping Kiev's overpriced hotels by renting an inexpensive apartment that was nevertheless nicer than any I had ever lived in back home. The front door of my building opened onto the bustling but cozy street of Zhitomirskaya, and it was an easy walk to Saint Sophia Square. There was also a nice terrace park where the young and hip of Kiev would gather in the late afternoon to throw Frisbees, play bongo drums, and drink beer in the glow of the sunset.

It all filled me with a churning panic. I just don't like being a clueless foreigner in a strange city where I've got no friends. I was also having trouble finding a portable radiation detector for my trip to the Exclusion Zone around Chernobyl. The detector would come in handy for measuring my radioactive exposure, with great precision, in units I wouldn't understand.

But Amazon didn't deliver overnight to Kiev, and so I was out of ideas. It was time to be resourceful—I had to get someone else to figure it out.

If journalism can teach us anything, it's that local people are a powerful tool to save us from our own fecklessness and incompetence. We call them *fixers*. In my case, I hired a capable young journalism school graduate called Olena. Skeptical at first, she soon realized that I was less interested in a simple rehash of the local disaster story than in exploring new touristic horizons, and she warmed to the concept. Olena set to work finding the radiation detector, calling one Chernobyl-related bureaucracy after another. To our surprise, nobody had any ideas. Even the government's Chernobyl ministry, Chernobylinterinform, was clueless. Measuring radiation didn't seem to be much of a priority among the citizens of Kiev. Maybe they just didn't want to think about what lay a short way upriver.

It's possible there is wisdom in such willful ignorance. The subject of radiation, after all, is so mysterious, and its units and measurement so confusing, that carrying around a little beeping gadget may not, in the end, leave you any better informed about your safety.

But every visitor to Chernobyl should have a working understanding of radiation and how it's measured. So let's review the basics. You can skip this section if you want, but you'll miss the part where I tell you the one weird old tip for repelling gamma rays.

Radiation, as far as tourists need be concerned, comes in three flavors: alpha, beta, and gamma. One source of radiation is unstable atoms—those same atoms that are so useful in building a nuclear core. In contrast to lighter, trustier elements like iron or helium, uncomfortably obese elements like uranium and plutonium are always looking for excuses to shed bits of themselves. That is to say, they are radioactive. These unstable elements will occasionally fart out things we call alpha or beta particles or gamma rays—the latter being the nasty stuff. This process—called *decay*—leaves the atom a bit smaller and sometimes with a different name, as it is alchemically transformed from one radioactive element into another.

Once in a while, an atom will suffer a complete breakdown and split in half. That's fission. After the split, particles and gamma rays spew off in all directions, and two atoms of a lighter element are left behind.

But we'll get to that. The point is, between decay and fission and other sources, there's radiation zipping around and through us all the time. There's the natural decay of Earth's atoms, and there are cosmic rays shooting down at us from outer space (you get a higher dose when you're up in an airplane), and then there's the X-ray your dentist gave you, and so on. You're getting irradiated all the time. But don't freak out yet. Although radiation can burn your skin, give you cancer, and disrupt the functioning of your very cells, it takes a lot of it.

And that's the problem. How much counts as a lot? It's hard even to predict how badly you'll get sunburned on a day trip to the beach, and that's with plain old solar radiation. In the nuclear case, your only hope of a clue is to have a radiation detector on hand.

Even with a detector, you're likely to remain confused by the bewildering array of terms and units with which radiation and radioactive dosage are measured. There are rads and rems, sieverts and grays, roentgens, curies and becquerels, around which buzz a swarm of attending coulombs, ergs, and joules. You might want to know the disintegration rate of a radioactive material, or its potential to ionize the air around it, or the amount of energy it can impart to solid matter, or the amount of energy it actually does impart to the living tissue of hapless organisms—such as Chernobyl tourists—and on and on.

And then it all depends on how quickly you get your dose. In this, radiation is analogous to certain other poisons, such as alcohol. A single shot of bourbon every weekend for a year is hardly dangerous. But fifty shots on a single night will kill you.

Finally, it matters which *part* of your body gets irradiated. Limb? Count yourself lucky. Guts? Not so much.

So it's no wonder that radiation is so mysterious and frightening, and that it features in the backstories of so many comic book monsters. It's invisible, deadly, cosmic, extremely confusing, and rides shotgun with the nuclear apocalypse. The stuff is just spooky, and if—like me—you're never going to have an intuitive understanding of its dosage and true risks, you might as well ease off on worrying about it so much. The purpose of the detector, then, is not to better understand the danger in your environment, but to gather up your anxiety and bundle it into a single number on a small digital readout, so you can carry your fear more efficiently.

Oh—and the tip for repelling gamma rays is that you can't.

\\\\\\\\\

Olena had a plan. "Let's go to Karavayevi Dachi," she said. "Electronics black market."

What Manhattan's Chinatown is to food, Kiev's Karavayevi Dachi is to electronics. It was early afternoon when we arrived. Metal stalls lined its alleys, roller-fronts thrown up to reveal jumbles of electronic components and devices. Men with rough, sun-cured faces sat at wooden folding tables strewn with vacuum tubes, transformers, electrical plugs, computer chips, adapters. The husks of car stereos hung in bunches, banana-like. It seemed doubtful that we would find a working radiation detector here among the tangled heaps of wires and transistors, and as we went from stall to stall, the conversations followed a pattern that always ended in "nyet."

One man claimed to have a detector at home that he would sell us for only 150 hryvnia—about thirty bucks. The catch was that it could only detect beta radiation. Forget it, dude. Any simpleton knows that beta particles—which can be blocked by regular clothes—are nowhere near as scary or stylish as gamma rays. We walked off to another part of the market, the annoyed vendor calling after us in protest, "But beta is the best!"

Another man had been eavesdropping and now approached us. He knew of a better place to find radiation detectors, he said, just a five-minute walk up the street. He would be happy to take us there. We left Karavayevi Dachi behind and made our way up a tree-lined street of brick apartment blocks.

Our guide's name was Volod. A middle-aged man with receding hair brushed straight back, he wore a beige coat over a striped beige shirt and beige jeans, and didn't seem to have much more of an idea than I did of where we were going. Our five-minute walk grew to fifteen and then twenty minutes, and I became progressively less convinced that we were detector-bound.

Striking up a conversation, we soon learned that Volod had been a communications officer in the Soviet army during the 1980s and had worked in the Exclusion Zone for two weeks, starting a month after the accident.

I asked him if he had received a liquidator's certificate. The "liquidators," in the creepy argot of the accident, were the thousands of workers, mostly soldiers, who had spent months razing villages to the ground and covering them with fresh earth, washing off roads, even chopping down and burying entire contaminated forests. The destroyed reactor had coated the landscape with radioactive dust and peppered it with actual chunks of nuclear fuel hurled clear by the explosion. It had been critical not to leave all that waste out in the open, where it could be tracked out of the zone or blown into the air by the wind. The liquidators' job was to clean it all up.

Many liquidators received high doses of radiation, and in consideration for their work, they were given special ID certificates that confirmed their status as veterans of the disaster cleanup. They were entitled to certain benefits, including special healthcare and preferential treatment in the housing system, but these benefits varied depending on each liquidator's dosage, on how soon after the accident he'd gotten there, and on how long he'd worked in the zone. A new mess was created—this time governmental. The system was riddled with loopholes, inconsistent in awarding benefits, and extravagant in its opportunities for corruption.

Volod told us he had not received a certificate. He hadn't been in the zone long enough, or early enough, he said. He didn't want to talk about it.

We were a good half hour from Karavayevi Dachi when he cried out. He had spotted the fabled detector store at last, among a small row of shops on the ground floor of an office building. We entered at a triumphant stride.

With its zealous air-conditioning and spotless tile floors, the store was a step up from the grungy maze of Karavayevi Dachi. Its offerings, though, were even more varied. Scanning the room, I saw shelves of videophones and security cameras next to displays of construction paper, coloring books, and crayons. Behind us, an entire section was given over to a plastic oasis of elaborate garden fountains cast in the shape of tree stumps. Between this and the Chernobyl Museum, I was beginning to discern a Ukrainian national genius for eclecticism.

And they sold radiation detectors. PADEKC, said the brand name on the box. NHDNKATOP PADNOAKTNBHOCTN. The device itself was a small, white plastic box with a digital readout and three round buttons. It looked like an early-model iPod, if iPods had been built by PADEKC. It was simple and stylish, perfect for hip, young professionals on the go in a nuclear disaster zone. Leonid—the salesman—assured me that it could measure not only gamma radiation but alpha and beta as well. (Leonid was a liar.)

He turned it on. "Russian made," he said. We crowded around. The unit beeped uncertainly a few times, then popped up a reading of 16. Sounded good to me. I coughed up far too many hryvnia and tossed the PADEKC in my backpack, and we went outside.

In front of the store, Volod asked for some money. I had been dreading his price.

"You should pay me vodka money," he said without irony. "A good bottle will cost about twelve hryvnia." He considered a dollar's worth of vodka decent pay for an hour's work. I handed him twenty hryvnia. As he started for the street, I asked him if he would tell us more about his time in Chernobyl.

He stopped and turned to us, suddenly taller.

"As a former Soviet officer, I cannot," he said. And then he wandered off to buy his vodka.

\\\\\\\\\\

The problem with Reactor No. 4 was not so much that its safety systems failed—although you could say they did—but that some of those systems had been disabled. Now, you could also argue that when you're running a thousand-megawatt nuclear reactor, you should never, ever disable any of its safeguards, but then . . . well, there's no *but*. You'd be right.

Those systems were disabled by an overzealous bunch of engineers who were eager to run some tests on the power plant and thought that they could do so without a safety net. On the evening of April 25, 1986, they began an experiment to see if the reactor's own electrical needs could be supplied by a freewheeling turbine in the event of a power outage. This is a little bit like stalling your car on the highway and trying to use its coasting momentum to run the AC. But in this case, it involved a three-stories-tall pile of nuclear fuel.

Over the course of several hours, the reactor failed to run hot enough for the experiment to proceed, so more and more control rods were pulled out of the core to juice the chain reaction up to a suitable level. Meanwhile, the flow of water through the core had dropped below normal levels, allowing more and more heat and steam to build up inside the reactor, a condition that made it dangerously volatile. The built-in systems to prevent all this from taking place were among the safeguards that had been shut off so the test could proceed.

In the wee hours of April 26, the operators noticed a spike in the heat and power coming from the reactor and realized that, if the control rods weren't reinserted immediately, the reactor would run out of control. It is assumed that they pushed the panic button to drop the control rods back into the core and stop the chain reaction—but panic was not enough. Not only were the rods too slow in sinking down into the reactor, but as they did so, they also displaced even more water, actually increasing the rate of reaction for a moment. The horrified engineers were powerless to stop it.

Within seconds, the power level in the core outstripped its normal operational level by a hundredfold. The water in the core exploded into superheated steam, blowing the two-thousand-ton reactor shield off the core. Moments later another explosion—possibly of steam, possibly of hydrogen, possibly an event called a nuclear *excursion*—punched a gaping hole in the top of the building. Bits of nuclear fuel rained down on the reactor complex and nearby landscape, setting the building and its surroundings on fire.

Inside the core, unknown to the panicking staff, the superheated blocks of graphite that formed the matrix of the reactor had also burst into flame, and the remaining nuclear fuel, completely uncontrolled, was melting into a radioactive lava that would burrow its way into the basement. All the while, radioactive smoke, dust, and steam spewed into the sky, a giant nuclear geyser that continued to erupt for days on end.

Before long, radiation sensors in Sweden were picking up the downwind contamination, and American spy satellites were focusing on the belching ruin of the reactor building, and the whole world was wondering exactly what the hell had happened in Chernobyl. In some ways, we still don't know.

\\\\\\\\\\

My pants made nylon swooshing sounds as I descended the musty stairwell of my apartment building. I had bought a tracksuit for my weekend trip to Chernobyl, which made me look like a Ukrainian jackass circa 1990, but it was disposable in case of contamination.

By chance, I was staying right across the street from the Chernobylinterinform, from which my trip to the zone would depart. This was in the time before Chernobyl tourism became officially sanctioned. In 2011, the policy was changed so that any yahoo could sign up for a tour through a travel agent—whereas in my day, that yahoo had to . . . sign up for a tour through a travel agent. I really don't know the difference, except that now the tours are officially offered to the public as tourism. Like most destinations after the word gets out, the place is probably ruined by now.

It was a beautiful day for a road trip, cloudless and faintly breezy. Nikolai, a lanky young driver for the Chernobyl authority, found a radio station playing insistent techno that suited his cheerful urgency with the accelerator, and we made our way out through the busy streets of northern Kiev. (Radiation level at the gas station: 20 microroentgens.) We followed the

Dnieper River north, until it wandered out of sight to the east. The road coasted through undulating farmland, bordered in stretches by lines of shady trees screening out the rising heat of the day. In our little blue station wagon, we plunged through villages, tearing past a boy idling on his bicycle, an old lady waddling along the road, a horse-drawn cart loaded with hay.

Soon there were no more villages, only countryside and thickening pine forests dotted with fire-warning signs. Compared with forest fires in the United States, which are disastrous mostly for their potential to destroy people's houses, a forest fire in the Chernobyl area carries added detriments. Trees and vegetation have incorporated the radionuclides into their structure, mistaking them for naturally occurring nutrients in the soil (one reason to shy away from produce grown near the zone). A forest fire here has the potential to release those captured radioactive particles back into the air and become a kind of nuclear event all its own. It's just one way in which the accident at Chernobyl has never really ended.

Less than two hours out from Kiev, we arrived at a checkpoint. A candy-striped bar blocked the road between two guardhouses. There were signs with a lot of exclamation points and radioactivity symbols. Nikolai and I stepped out of the car and I gave my passport to the approaching guard. He wore a blue-gray camouflage uniform, a cap bearing the Ukrainian trident, and a little film badge dosimeter on his chest, to measure his cumulative exposure while in the area. I should have asked him where I could get one of those.

After a cursory search, we hopped back into the car. The barricade rose, Nikolai gunned the engine, and we left the checkpoint, traveling onward through the forest, down the middle of a sun-dappled road that no longer had a center line.

\\\\\\\\\

We had entered the Exclusion Zone.

At the Chernobylinterinform administrative building, in the town of Chernobyl—nearly ten miles distant from the reactor itself—we met Dennis, my escort. Standing at the top of the steps to the low, yellow building, Dennis matched the quasi-military vibe of the zone. He was in his mid-twenties, with an early baldness made irrelevant by a crew cut, and wore combat boots and a camouflage jacket and pants. The look was completed—and the martial spell broken—by a black sleeveless T-shirt printed with the image of a football helmet, around which swirled a cloud of English words. A pair of wraparound sunglasses hid his eyes.

"First is the briefing," he said coolly. "This is upstairs." And with that he walked back into the building.

The briefing room was a long, airy space, its walls hung with photographs and maps. A wooden table surrounded by a dozen chairs dominated the center of the room. The floor was covered with an undulating adhesive liner printed to look like wood paneling. Must make for easy cleanup, I thought, in case anyone tracks in a little cesium.

Dennis and I were alone. The summer season hadn't picked up yet. He retrieved a gigantic wooden pointer from the corner, and we approached a large topographic map on the front wall. He began diagramming our itinerary using his tree limb of a pointer, though the map was mere inches in front of us.

"We are here. Chernobyl," he said, and tapped on the map. "We will drive to Kolachi. Buried village." He tapped again. "Then to Red Forest. This is most radioactive point today." He looked at me for emphasis. He was still wearing his sunglasses.

Turning back to the map, he continued. "From here we will go to Pripyat. This is deserted city. Then we can approach reactor to one hundred and fifty meters."

It was the standard itinerary, allowing visitors to inhabit their preconceptions of Chernobyl as a scene of disaster and fear—but without actually straying off the beaten path or risking contamination. This was, after all, what most people wanted. But I hadn't come all this way only to wallow in post-nuclear paranoia. I was here to enjoy the place, and this was the moment to make it happen.

"Is there any way . . ." How to put it? "Is there any way we could go canoeing?"

Dennis regarded me blankly from behind his shades. In their silvery lenses, I could see the reflection of someone who looked like me, with an expression on his face that said, *Yes. I am an idiot.*

"This is not possible," said Dennis.

"Well, if there's any way to get on the water, or maybe visit the local fishing hole, I'm happy to sacrifice part of our planned itinerary."

The conference room was quiet. The shadow of a grimace passed across Dennis's face. "This is. Not possible," he said, without emotion. He was proving witheringly immune to what I had hoped would be my contagious enthusiasm. But it's at moments like this, when you're trying to take your vacation in a militarily controlled nuclear disaster zone—for which, I might add, there is no proper guidebook—that you must be more than normally willing to expose yourself as a fool in the service of your goals. I laid my cards on the table.

"Look. Let's say I wanted to go for a boat ride with some friends some-where in the zone," I said. "Just theoretically speaking, where would we go? I mean, where are the really *nice* spots?"

A faint crease had developed in the dome of Dennis's head.

I pressed on, telling him that I was trying to approach this not so much as a journalist or a researcher, but as a tourist. As a *visitor*. Where, for instance, could I find a good picnic spot in the Exclusion Zone? Where did he himself go on a slow day? And if it wasn't possible in the zone, what would be the next best thing? I pointed to Strakholissya, just outside the zone, a town that I had identified while poring over a map the night before. What about that?

"Yes, this is nice place," said Dennis. "You can go fishing here."

I was making progress. Fishing?

"Yes," said Dennis, gaining speed, "but this place is better." He pointed to Teremtsi, a tiny spot nestled among a bunch of river islands deep inside the zone. "This is a good place for fishing," he said. "I went once. Mostly I go there to collect mushrooms."

I stared. Mushrooms, because they collect and concentrate the radionuclides in the soil, are supposed to be the last thing you should eat in the affected area. And Dennis gathered them in the heart of the Exclusion Zone.

"You collect mushrooms? And you eat them?" There was awe in my voice.

"Yes, this is clean area, I know. This is no problem."

I couldn't believe my luck. A total newbie, I was already teamed up with a guy who used the zone as his own mushroom patch and trout stream. I wanted to abandon our itinerary. Who needs to see a destroyed nuclear reactor when you can go fishing just downriver?

Don't think I didn't beg. But Dennis was far too professional to chuck the official program—with all its approved paperwork, stamped and signed in duplicate for each checkpoint—just because some half-witted foreigner said pretty please. But this time, there was a moment's hesitation. "This is, um. Not possible," he said, getting back on script. But I saw the hint of a smile on his face as he turned away from the map.

The briefing continued down the side wall of the room, from one diagram to the next. There were a pair of maps showing the distribution of contamination by radioactive isotopes of cesium and strontium in the zone. The contamination is wildly uneven, depending on where the radioactive debris fell immediately after the explosion, and on the wind and rain in the days and weeks that followed, when the open reactor core was spewing a steady stream of radioactive smoke and particles into the air. The weather of those following weeks is inscribed on the ground, in contamination. The maps showed the distribution, color-coded in shades of red and brown, a misshapen starfish with its heart anchored over the reactor.

The radiation level at any given spot also varies over time, although based on what, I'm not quite sure. So there are limits set for what is considered normal, just as there are in any city. Dennis told me that the standard in the town of Chernobyl was 80 microroentgens per hour. In Kiev, it was 50. (It's about the same in New York, where background radiation alone gets you about 40 micros per hour.)

"In the last month, I measured 75 micros at different places in the town," he said. Chernobyl was pushing the limit. But I was unclear what the standards for radiation levels really meant. In Kiev, for example, what difference did it make that the standard was 50 micros instead of 80, or 100?

"It means," said Dennis, "there would be panic in Kiev if the reading was 51." He thought people in Kiev were a little paranoid about contamination. "Just yesterday, some journalists called, saying they had heard there was a release of radioactive dust at the reactor," he said. "I told them I had just been down to the reactor, and there was no problem."

We had come to the end of the briefing. Dennis paused in front of the last photograph, which showed a large outdoor sculpture. Two angular gray columns held a slender crucifix aloft, like a pair of gigantic tweezers holding a diamond up for inspection. Below them, half a dozen life-size figures lugged fire hoses and Geiger counters toward a replica of the reactor's cooling tower.

It was the firemen's memorial. In the hours immediately following the explosion, the firemen of Pripyat had responded to the fire that still burned in the reactor building, and had kept it from spreading to the adjacent reactor. Unaware at first that the core had even been breached, they received appalling doses of radiation and began dying within days.

Dennis turned to me, expressionless behind his shades. The giant pointer tapped gently on the photo of the memorial. "If not for those firemen," he said, "we would have an eight-hundred-kilometer zone, instead of thirty."

\\\\\\\\\\\

Nikolai was waiting in the parking lot, smoking a cigarette. "We work until five o'clock," Dennis said. "So we'll have lunch at half past four. Then at ten to five is football. It is the most important game." He was talking about the World Cup. Ukraine's soccer team had qualified for the first time, and tonight was the critical elimination match against Tunisia. Fanciful images of a raucous Chernobyl sports bar danced through my mind. I told Dennis it sounded like a great plan.

He rode shotgun, while I sat in back. A few hundred yards beyond the firemen's memorial, Nikolai pulled into a small gravel parking lot and jumped out of the car to buy a bottle of beer and an ice cream bar. There was a convenience store in the zone.

Within a few minutes we reached the checkpoint for the ten-kilometer limit, which encompasses the most-contaminated areas. The car barely stopped as Dennis handed a sheet of paperwork through the window to the waiting guard. He folded the rest of our permission slips and tucked them into the car's sun visor for later.

The air that streamed through the car's open windows was warm and sweet, a valentine from the verdant countryside that surrounded us. It felt as though we were just three guys out for a pleasant drive in the country—which was more or less the truth. Dennis and Nikolai traded jokes and gossip in Russkrainian. "We're talking about the other guide," said Dennis. "He's on vacation." It seemed there were no more than a handful of Chernobylinterinform guides. It only added to the sense that I had found a traveler's dream: an entire region that—although badly contaminated—was beautiful, interesting, and as yet unmolested by hordes of other visitors.

My thoughts were interrupted by a loud electronic beep. My radiation detector had turned itself on—funny, that—and now that there was actually some radiation to detect (a still-modest 30 micros), it had begun to speak out with an annoying, electric bleat that in no way matched the PADEKC's smooth iPod-from-Moscow look. There was a reason, I now realized, that this detector looked like something you might take to the gym instead of to a nuclear accident site: It was designed for the anxious pockets of people who thought 30 micros were worth worrying about.

In the front seat, Dennis had produced his own detector, a brick-size box of tan plastic fronted by a metal faceplate. Little black switches and cryptic symbols in Cyrillic and Greek adorned its surface. I was jealous. It seemed there was no kind of radiation it couldn't detect, and it probably got shortwave radio, too. Its design was the height of gamma chic: slightly clunky, industrially built, understatedly cryptic, and pleasingly retro. What really sold me was its beep. Unlike the fretful blurts of the PADEKC, the beeps of this pro model were restrained, almost musical. It sounded like a cricket, vigilantly noting for the record that you were currently under the bombardment of this many beta particles, or that many gamma rays. It was a detector made for someone who accepts some radiation as a fact of daily life, and who doesn't want to lose focus by being reminded of it too loudly. Someone who is perhaps even something of a connoisseur of radiation levels. Someone like Dennis.

The car stopped. Dennis pointed out the window to a large mound among the trees. "This is Kolachi," he said. A pair of metal warning signs stood crookedly in the tall grass. That was all that remained. The village had been so contaminated by the accident that it was not only evacuated but also leveled and buried. There were many such villages. Dennis held his meter out the window: 56 micros. It was my first time above Kiev-panic levels.

Leaving Kolachi behind us, we passed by some old high-tension electrical wires—presumably part of the system that had, until recently, carried electricity from the undamaged reactors of the Chernobyl complex, which had continued to function even after the accident. The Soviet and then Ukrainian governments kept the other three reactors running into the 1990s, and only shut the last one down in 2000. Of the thousands of people still reporting to work in the Exclusion Zone, the great majority are employed in the decommissioning of those reactors.

On the seat beside me, the PADEKC was getting insistent. I apologized for the racket, feeling the same embarrassment as when your cellphone rings in a quiet theater. I tried to put the damn thing on vibrate, but all I could manage was to get lost in its impenetrable Russian menus. When we took the turnoff for Pripyat, it began to freak out in earnest. The reading ascended quickly from 50 micros through the 60s and the 80s, and into the low 100s. The beeping increased in pace, in a way I could only find vaguely alarming. Nikolai glanced back at me, unconcerned, but wondering what my little meter was making such a fuss about.

We were crossing through the Red Forest. Named for the color its trees had turned when they were killed off by a particularly bad dose of contamination, the Red Forest was cut down and buried in place, becoming what must be the world's largest radioactive compost heap. Back in the briefing room, Dennis had warned me that we would experience our highest exposure while passing through this area, which had since been replanted with a grove of pine trees, themselves stunted by the radiation.

As we rounded a bend, Dennis again held his meter aloft outside the passenger window. It began chirping merrily. Meanwhile the PADEKC was going nuts. In Kiev, Leonid had told me the upper limit on the unit was 300 microroentgens, but it now spiked from the mid-100s directly to 361. The car filled with our detectors' escalating beeps, which quickly coalesced into a single shrill tone that was painfully reminiscent of the flatlining heart monitor you hear on hospital TV shows.

Dennis's meter topped out at 1,300 micros, about thirty times the background radiation in New York City. He twisted around in his seat to face me. "Yesterday it was up to 2,000," he said. There was a hint of apology in his voice. Perhaps he was worried I might feel shortchanged for having received less than the maximum possible exposure from a Red Forest drive-by, as if I had come to Nepal to see Mount Everest, only to find it obscured by clouds.

Over a bridge lined with rusted street lamps and ruined guardrails, Nikolai slowed the car to weave between the potholes dotting the roadway. At the bottom of the bridge's far slope, we reached another checkpoint. Dennis adroitly snatched the next leaf of paperwork out of the stack and tucked it into the waiting hand of the guard. The sign at the checkpoint read PRIPYAT.

\\\\\\\\\\\

Even more than the reactor itself, Pripyat is the centerpiece of any day trip to the Exclusion Zone. Before 1986, it was a city of nearly fifty thousand people, devoted almost entirely to running the four nuclear reactors that sat just down the road and to building the additional reactors that were to be added to the complex. At the time of the meltdown, Reactors Nos. 5 and 6 were nearing completion, and a further six reactors were planned, making the neighborhood a one-stop shop for the area's nuclear energy needs.

It didn't take long for the residents of Pripyat to realize there had been an accident. Anyone looking south from the upper stories of Pripyat's tall apartment buildings could have seen smoke belching from the maw of the destroyed reactor building some two kilometers distant. What they didn't know was that it was no ordinary fire.

The city was bathed in radiation, but the residents remained uninformed. They continued about their business for more than a day, while the government scrambled to contain the accident. Finally, at noon on April 27, nearly a day and a half after the explosion, the authorities announced their decision to evacuate the city.

You can say what you like about the Soviet government, which killed and exiled millions of its own people and repressed most of the rest. But you have to concede that when they put their minds to it, they really knew how to get a place evacuated. When at last the order was given, it took only hours for this city of fifty thousand people to become a ghost town. The evacuation was broadened over the following days to include more than a hundred thousand people. Ultimately, more than three hundred thousand were displaced.

Pripyat sat empty. In the months and years following the evacuation, it was looted and vandalized by people who were obviously unconcerned by the radioactive nature of their spoils, whether televisions for their own use or metal items to be sold as scrap. The evacuation and the looting turned Pripyat into what it is today: the world's most genuinely post-apocalyptic city.

In spite of what you might have seen in the movies, though, things can actually be pretty nice after an apocalypse—if a bit scarce in terms of human beings. The road that led us into Pripyat from the south was lined with bushes speckled with small white blossoms, the air thick with the smell of flowers. The vista opened up as we reached the center of town, allowing a view of the buildings around us. Dennis and I clambered out in the middle of an intersection, and Nikolai motored off down a side road to find a nice spot to sit and drink the beer he'd bought earlier.

The day was hot and sunny. The ghostly city surrounded us, the buildings of downtown looming up from behind scattered poplar trees. Behind us rose a ten-story apartment block. Its pink and white plaster facade was falling off in patches, revealing the rough brickwork of the walls underneath. More apartment blocks stood along the road to the left, some of them crowned with large, Soviet hammer-and-sickle insignia that must once have lit up at night.

We walked toward the town plaza, following a path that had once been a sidewalk but was now a buckling concrete track invaded by weeds and grass. Dennis lit a cigarette and looked up as he took a long drag. A gentle breeze pushed a herd of little clouds across the sky. Birds flitted by.

The plaza was bordered on three sides by large buildings. To the right, a defunct neon sign announced the Hotel Polissia, seven stories of square, gaping windows. From where we stood, more than twenty years of looting and abandonment had not significantly worsened the stark, unforgiving aspect of the hotel's architecture. A few hardy shrubs even peeked from among the freestanding letters of the roof sign. It's amazing where things will grow when people stop all their weeding.

Between drags on his cigarette, Dennis answered my questions with the jaded economy of someone who had been to this spot a thousand times. "What's that?" I said, pointing at the building to the left of the hotel.

"Culture palace," he said.

"What's a culture palace?"

"Discos." Another drag. "Movies."

To our left was a blocky building with a sign reading РЕСТОРАН. Using my nascent Cyrillic, I decoded this as RESTAURANT. I pointed to a low-slung gallery that jutted from its side.

"What was there?"

Dennis looked up and removed the cigarette from his mouth.

"Shops."

The plaza where we stood was gradually surrendering itself to shrubs and moss. Vegetation spilled over its borders and crept along its seams. A set of low, crumbling stairs led up from the plaza's lower level, purple wildflowers and a few tree saplings poking out from the cracks.

"Don't step on the moss," Dennis ordered as we walked up the mossy stairs from the mossy lower level to the mossy upper level.

"Why's that?" I asked, and hoped he hadn't seen the contorted tap dance of my reaction.

"The moss . . . concentrates the radiation," he said, and tossed his cigarette butt on the ground. The same could have been said for the mushrooms he had freely admitted to gathering in the zone, but I didn't bother to point it out.

I stopped to take a picture. Dennis dangled another cigarette from his mouth and posed on the concrete path, the pectopah in the distance. Behind his sunglasses, he could have been the bassist for a Ukrainian rock band. FOOTBALL, said the writing on his sleeveless black T-shirt. SYNTHETIC NATURE. He held up his detector for the camera to see. It read 120. But what did 120 microroentgens mean on a sunny day? More than a little. Less than a lot. Panic in Kiev.

Dennis wandered away along the side of the plaza, his detector in a lazy warble. I lingered in front of the gutted pectopah. There was nothing left but a shell of cracked concrete and twisted metal. I tried to imagine the plaza before the accident, when it had been the center of a living city. A place to meet a friend after work, maybe. Somewhere to have a cup of bad coffee. What was it like to have your entire town evacuated in three hours? To lose not only your house or apartment but also your workplace, your friends, your entire environment? I tried to imagine the terror of that day.

But in the peace that reigned over present-day Pripyat, it was difficult. I closed my eyes and felt the sun on my face. The trees and grass rustled in the wind. Insects buzzed past on their way to somewhere else. I heard the easy cacophony of the birds. And as Dennis made his way down the plaza, the chirping of his dosimeter dissolved into the birdsong, becoming just another note in nature's symphony.

I caught up with him at the far corner of the shops, and we headed around back to visit the amusement park. As we walked, I asked Dennis how he had gotten his job. Leading guided tours through the world's most radioactive outdoor environment didn't seem like a gig you would find on Craigslist. And Dennis had started early; at twenty-six, he had already been working for the Chernobyl authority for three years, alternating every two weeks between the zone and Kiev to keep his radioactive dose under the permitted limit. He told me that originally he had worked only in the Kiev office, before getting transferred to the zone. "I asked for it. I wanted to do it instead of sit in front of a computer," he said, and took a swig of water. "And most people don't work at all, if the computer has Internet." Here was someone who believed that boredom was worse for you than radiation.

He went on, recounting how he had read about a doctor who argued that constant, low-level doses of radiation were actually good for you. "The people who didn't leave the zone after the accident lived better," he said, referring to the several hundred aging squatters who have been allowed to live, semi-legally, in their houses in the zone. "This doctor said they had adapted to the radiation and would die within fifteen years if they suddenly leave, but could live to a hundred if they stay."

I had heard similar claims before, and I was doubtful. There probably were health benefits for zone squatters, but surely they came from living in a little cottage in the countryside, where they grew their own (albeit contaminated) vegetables and breathed clean (if radioactive) air, instead of being evacuated to a crappy apartment in Kiev. I suggested this to Dennis, that perhaps around here, quality of life just trumped radiation dose. He shrugged. But not everyone in Ukraine was as casual as he was about radiation. He later told me how, whenever he would visit his sister in Kiev, she would make him leave his boots outside.

The amusement park is Pripyat's iconic feature, an end-times Coney Island, with a broad paved area surrounded by rides and attractions that are slowly being overcome by rust and weeds. Dennis was more interested in the moss. He was a collector of hotspots, and around here the moss had all the action. Near the ruined bumper car pavilion, he waited for a reading before picking his radiation meter up from a mossy spot on the ground.

"One point five mili," he said, wiping the meter's backplate on his fatigues.

We left the amusement park and walked down the street, past the post office, past a low building that Dennis said was a technical school, past more apartment blocks. Turning off the road, we scurried through a large concrete arch attached to another building; Dennis eyed the unstable structure warily as we passed underneath.

We continued through the rear courtyard of the building and into an overgrown area beyond. The warm shade of the forest was alive with the hum of bees. As we walked, I pushed aside branches and squeezed between bushes that grew in our way. At the end of the narrow path was a two-story building made with pink brick set in a vertical pattern.

"Kindergarten number seven," said Dennis.

If you have been insufficiently sobered by the sight of a deserted city, Kindergarten No. 7 will do the trick. We came through a dank stairwell into a long, spacious playroom with tall windows on one side, their glass long since smashed out. Thick fronds of peeling, sky-blue paint curled from the walls. What had been left behind by the looters—or shall we call them the first tourists?—was strewn on the floor and coated with twenty years of dust from the slowly disintegrating ceiling. Mosquitoes made lazy spirals through the humid air.

The door was torn off its hinges. Next to it lay piles of orange play blocks and a mound of papers printed with colorful illustrations—marching elephants, rosy-cheeked little Soviet children. A gray plastic teddy bear, its face pushed into the back of its hollow head, sat on a moldering pyre of Russian learn-to-read posters. I recognized the Cyrillic letter *b*.

Dennis was by the windows with his detector. "Eighty," he said. He walked to the far wall. "Five."

A toy car with a yellow plastic seat just large enough for a single child was parked in the middle of the room. It was missing its wheels and its windshield. Even it had been stripped for parts. On the floor next to it was a child-size gas mask.

Stepping around pools of stagnant water, we made our way out through the stairwell, pausing in front of some black-and-white photographs still hanging on the wall. In them, children played and did exercises in a tidy classroom. With a gnawing temporal vertigo, I felt the pictures snap into familiarity: It was the same room. The destroyed room we had just left. And the toys the children were playing with in the photographs were the same toys we had seen just now, fossilized in dust.

Nikolai picked us up on the street, the car appearing out of nowhere, and we left Pripyat in silence.

The classroom lingered in my mind. I had come to the Exclusion Zone to witness its unexpected and riotous efflorescence, and there was something joyous in the sight of nature rushing into an unpeopled world. But it was a garden fed with suffering. Although the meltdown in Chernobyl was no death sentence for the people of Pripyat—and although most of the children who attended Kindergarten No. 7 are probably alive and well today—at the bare minimum it displaced and terrorized hundreds of thousands of people, and threw a pall of doubt over their health, a sickening uncertainty that will haunt the region for at least a lifetime. In this, the verdant bubble of the zone was unlike any other oasis in the world. It had been wrenched into existence, with violence. Something had created it.

On the far side of the bridge out of Pripyat, we coasted to a stop. Dennis turned to me. "Perhaps you would like to take a picture," he said. I was confused. Why here? But then my eyes wandered up to the horizon, and for the first time, I saw the reactor in person.

It hunkered in the distance, perhaps a mile away, its latticed cooling tower rising over a nasty confusion of buttressed metal walls. The Sarcophagus. Officially known as the Shelter Object, it had been built to contain the shattered reactor. Floating over an expanse of low forest, it had a strange and massive presence. It could have been a crashed spaceship.

By the time we reached the reactor complex, the day had turned itself inside out. We had heard thunder rumbling in the southeast only moments after I'd first seen the Shelter Object. Now a thick lid of clouds had slid over the sky, and heavy raindrops were striking the car's metal roof. Our surroundings were similarly changed, overtaken by forbidding expanses of concrete and clusters of squat buildings—the infrastructure for maintaining the reactor building. Through the car's streaming windshield, I saw a dented metal gate blocking our way and a pair of concrete walls haloed with messy helixes of barbed wire.

On the other side of it all, attended by several spindly yellow construction cranes, was the Shelter Object. I was struck again by its great size. The interlocking metal walls rose in a colossal vault nearly two hundred feet tall, battleship gray streaked with rust, supported on one side by tall, thin buttresses and on another by the giant, blocky steps of the so-called Cascade Wall. Pipes and bits of scaffolding clung to its battlements, whose flat surfaces were interrupted by a grid of massive metal studs. Catwalks traced the edges of its multiple roofs, and a series of tall, shadowed alcoves notched the top of the north wall, like portals from which giant archers might rain arrows down on the countryside.

I had envisioned this moment differently. Visiting the reactor building, I had assumed, would not be fundamentally different from visiting the Eiffel Tower or the Taj Mahal. But those thoughts vanished under the growing thunderstorm. Instead, I felt an unexpected, visceral repulsion. It was obscene. This *thing*. A monument to brutality, a madman's castle under siege from within itself. And it lived. It radiated danger and fear. It had warped the land for miles around, creating its own environment, breathing the Exclusion Zone to life.

There was a visitor center. No gift shop, but there were diagrams and photographs, and an excellent scale model of the Shelter Object. I was met by Julia, a serious woman in her forties who gave me a quick handshake before loosing a torrent of information about the accident and the reactor building. Much of it I already knew, but it took on fresh weight in the aggressive Ukrainian accent of an Exclusion Zone bureaucrat. The visitor center's picture window gave the lecture additional dramatic punch. Through it we had the world's best, closest view of the ever-more-menacing Shelter Object, now crowned with forks of lightning. And in case that wasn't enough, there was an electronic readout above the window that measured our radioactive exposure—138 micros at the moment.

"Sarcophagus took two hundred six days to *construct*," said Julia. "Radiation levels at north side of building after *accident* reached 2,000 rem per hour." She waved her hand over the model, a perfect replica, two or three feet tall. "On top of *building* they reached 3,000 rem per hour. This is appalling *level*. These are area where firemen were working." I felt a little sick. Even several hundred rem can be fatal, and the first responders to the Chernobyl accident received many times that for every hour they spent on the building.

Thunder rattled the window. With practiced ease, Julia swung open the hinged front wall of the model to reveal a cross-section of the interior, its wreckage recreated in painstaking detail. With the actual building visible just out the window to the right, the model allowed an intuitive understanding of the gargantuan scale of the reactor—and of the accident that had destroyed it. Flicking my eyes back and forth, it was as if I could see right through the walls of the Shelter Object and into the building's guts. As Julia continued to reel off facts and figures, she lifted the roof off the turbine hall with the tips of her fingers, a colossal June Cleaver demonstrating how to use an Olympian piece of Tupperware.

The destruction inside was complete. The core's radiation shield, a two-thousand-ton plug of lead that had been blown into the air by the explosion, had landed on its side, and now hung precariously at the top of the core. The core itself was the size of a small building, a thick bucket standing several stories tall. It felt impossible to understand the power embodied in such a machine. A quarter ounce of nuclear fuel holds nearly as much energy as a ton of coal; the core had held more than a hundred thousand times that much.

Now, though, it was empty. Some of the fuel—nobody knows exactly how much—was ejected in the explosion and subsequent fire. The rest melted through the floor of the reactor, a nuclear lava flow that spilled into the lower floors and basements of the reactor building, where it still sits, unapproachably radioactive. Julia pointed to a photograph on the wall that showed some of the lava, a cracked cylinder with a flaring, globular base. "This is elephant *foot*. Is most famous portion of nuclear lava, in basement of building." She turned back to the model and indicated a number of tiny flags planted inside the core and around the building. "These are temperature and radioactivity sensors," she said. "They have been placed by Chernobyl workers."

I was incredulous. People had actually gone into the reactor core?

Julia nodded. "Yes. Duty cycle is fifteen minutes."

The idea of rappelling into the empty core made me dizzy. Julia went on, cataloging the Shelter Object's many problems. Its walls are riddled with gaps and small cracks; if any of the corroding wreckage inside the building shifts or falls, it may spew plumes of radioactive dust into the air outside. In the meantime, the gaps in the walls have allowed hundreds of gallons of rainwater in, water that has presumably trickled down through the building and created a kind of radioactive tea that may in turn seep into the ground-water.

Perhaps one of the worst parts of the situation, Julia offered brightly, was simply that nobody knew exactly how much nuclear material was inside the building, or just where it was, or what it was up to. Some scientists have even wondered if the trickling rainwater might be leaching impurities out of the solidified nuclear lava, slowly refining it. If this is true, it means that the fuel might one day reach sufficient purity for the chain reaction to start up again on its own, creating an uncontrolled nuclear campfire in the basement of the building. And even if that doesn't happen, the entire Shelter Object might just fall in on itself anyway. The west wall, supported by parts of the rotting interior structure, had shifted recently, taking its first small step toward a possible collapse.

I was ready to leave. Beneath the thunder rumbling outside, I imagined I heard a low throbbing sound coming from the reactor building. But Julia wasn't quite finished. She was telling me about the future of the Shelter Object. Because its sheltering will essentially never be done, it's impossible to dismantle it and replace it with something better. So first they're going to stabilize the thing, buttressing its buttresses and supporting its supports. And then—what else?—they're going to build a shelter for the Shelter Object. They call it the New Safe Confinement.

"New Safe Confinement won't just be a shelter," Julia intoned. "It will be a technological *complex*." She pointed to some conceptual drawings of the New Safe Confinement; they showed a tall arc of smooth concrete that soared over the whole mess with the same geometric élan as the St. Louis Arch. Robotic cranes will hang from its interior, in order to maintain the Shelter Object as it continues to decay. The New Safe Confinement, if it's actually built, is intended to last 150 years. The reactor building, though, will be dangerous for millennia. So maybe there will one day be a shelter for the shelter for the Shelter Object, and then a shelter for that, and we will continue down the generations, building—shell by shell—a nest of giant, radioactive Russian dolls.

Dennis appeared at the door—windswept and wet with rain, but still wearing his shades—and beckoned for me. Julia walked me out, talking continuously about the lack of funding for the New Safe Confinement, or even for the preliminary stabilization to keep the Shelter Object from collapsing in a heap. She emphasized that Ukraine needed international help for this, perhaps hoping that I would pass the message along to the White House or the United Nations. Chernobyl was the responsibility of the entire world, she said. Besides, Ukraine was too broke.

Emerging into the storm, Dennis shouted, "Here you can take a photograph, and let's go!" Pictures weren't allowed from inside the visitor center, not that I had felt like taking one. I turned into the wind and snapped a single, rain-spattered photograph of the Shelter Object before diving into the waiting car. Nikolai floored it.

As quickly as it had begun, the storm faded. The clouds broke as we passed the half-built forms of Reactors Nos. 5 and 6. The sun came out. A spectral curtain of steam rose from the road. Laughing at a comment from Nikolai, Dennis pointed to the vapor curling off the asphalt. "We're joking that now you can see the radiation," he said.

At Dennis's direction, Nikolai veered left and we catapulted up a gradual slope and onto a long, deserted bridge that spanned the river. This was the Pripyat River, which runs right past Pripyat and the Chernobyl reactor, and into which the cooling channel from the nuclear reactor drains. The Pripyat also empties into the Dnieper River, which runs through Kiev and is the backbone of Ukraine's most important watershed. You might call it the Ukrainian Mississippi, except there hasn't been a meltdown in Minneapolis yet.

Dennis had made this stop, I think, as a concession to my pleas for a tour of the zone's "nice spots." Nikolai killed the engine and we got out of the car and walked across the deserted road to the north side of the bridge. The river stretched away toward the power plant, a miniature in the distance. Dennis and Nikolai lit cigarettes and we leaned on the guardrail, staring out at the view. The wide, coffee-colored water of the river, gently iridescent with shafts of warm sunlight, rippled against a border of marshy grass and tall reeds. Beyond the tiny shapes of the cooling tower and reactor buildings, a forest of grumbling thunderheads retreated over the horizon. Peace descended again on the zone. The official part of the tour was over.

\\\\\\\\\\\

At headquarters, Dennis and I ate quickly and in good style. The dining room was air-conditioned (the remote control for the AC looked a lot like my radiation detector), the table was covered with an embroidered table-cloth, and the meal was multicourse, with plates of meats and cheese and vegetables (not local). For the first time, Dennis took off his sunglasses. He seemed uneasy with his eyes exposed to the light, and we sat stiffly at the table, trading snippets of conversation. Maybe he was worried about missing the start of the soccer game. As soon as I told him he didn't have to wait, he excused himself and headed upstairs.

The game had started by the time I joined him. What I had hoped would be a raucous gathering of soccer-crazed zone workers was actually a small, somber party of five people: only Dennis, Nikolai, a pair of tired, middle-aged secretaries from the Chernobyl authority office, and me. We were well provisioned, at least, with a generous spread of vodka, cognac, cola, and some kind of pickled fish. The game was scoreless into the second half, but we found moments to toast: a good save here, a near miss there. We would hold our glasses up, wait for a few words in Ukrainian from Dennis or Nikolai, and then drink. The secretaries glared at me meaningfully before each slug of vodka: the spirit of inclusion, I chose to think.

Finally Ukraine scored on a dubious penalty kick. The remaining minutes ticked away, and the game ended 1–0. Ukraine would be advancing to the next round. Nikolai pounded on the table in celebration while Dennis poured out another round. He looked me in the eye, our glasses raised.

"To victory," he said.

\\\\\\\\\\

Afterward, Dennis and I went for a walk, clutching liters of beer we had bought at the corner store. It was a beautiful Friday evening, still warm with lingering sunlight, and the town was quiet. I suppose the place is always quiet. The only other person in sight was Lenin, standing alone on a low concrete platform, his hand in his pocket, looking like he was waiting for the bus.

A car passed in the distance, and we hid our beer bottles inside our coats. "We are not supposed to have beer outdoors in the town," Dennis said. "If it is police, they can get angry." For a moment, I felt like a bored teenager in a too-small town, with nothing to do on a Friday night but wander the streets and get drunk. Maybe there's a reason the Exclusion Zone is also called the Zone of Alienation.

Across the street from Lenin, next to the church, was the recreation center. Dennis told me there was a Ping-Pong table inside, but that the place was closed for the weekend. First no canoeing or mushroom gathering, and now no Ping-Pong? These people had a thing or two to learn about hospitality.

"Come, I can show you the nice spots in town," Dennis said. We strolled to the edge of town and then down an overgrown dirt road toward the water. Now off the clock, Dennis had dropped the forced, semi-military formality of his guide persona and was enjoying himself. He pointed at the thick overgrowth spilling into the road. "There could be wild boar here," he said. "They like to hide in bushes like these. Sometimes the mother boars leap out of the bushes and charge. If this happens, you must climb something very tall, like this—" He pointed at one of the tall, concrete utility poles that lined the road.

I looked at it doubtfully. "I don't think I could climb that."

Dennis took a swig of his beer and smiled. "If the wild boar is charging, you learn fast."

At the riverbank, we stopped and stared out at the sunset, the surface of the water glassy and still. I wondered idly if the giant mosquitoes swirling around us were mutants, or if we might see a three-eyed fish. A few mutants would add such panache to the zone. But the closest you'll come are the deformed, runty trees of the Red Forest and some unspectacular abnormalities in bird coloring, in the litter size of the wild boar, in who knows what else. The point is there are no two-headed dogs.

The world thinks of Chernobyl as a place where humankind had overwhelmed and destroyed nature. The phrase "dead zone" still gets tossed around. But this was nowhere more obviously untrue than here, watching the sunset, my entire horizon a quiet rhapsody of water, sun, and trees. Paradoxically, perversely, the accident may actually have been good for this environment. The radiation—while not exactly healthy for any organism—has been so effective at keeping humans away that Chernobyl has gone back to nature, a great, unplanned experiment in conservation by way of pollution. For decades, wildness has been reclaiming the place, growing in where civilization would have pushed it back, reoccupying the space once reserved for people.

If the zone had become a giant, radioactive national park, then Dennis was the Boy Scout in love with it. As we walked back to town, birdsong filling the air, he told me about the scientists and researchers who came to the zone to study the wildlife. His pride was obvious. Species of birds not seen in the region for decades had been popping up there, he said. Ecologists had even chosen it as a place to reintroduce an endangered species of wild horse. And everywhere I had gone, except for the reactor complex itself, I had seen nature running riot. Despite the radiation—indeed because of it—Chernobyl had effectively become the largest wildlife preserve in Ukraine, perhaps in all of Europe.

It is a turn of events that highlights a certain human arrogance about our destructive powers. It is only hubris to imagine that we can destroy nature, or the world. It is the mirror image of the industrialist's egotistical desire to exploit and control it. And it is true that we can kill off continents of forest and destroy species by the thousands, and even wreak climate change. But once we're gone, the rest of nature will rush on, as it has after so many other cataclysms, growing over and through and out of us. The apocalypse we can create is for ourselves and for our cousins, but not for life on Earth.

We headed back by a different route, cutting through the town's World War II memorial, an arcade of pillars tucked into the woods. The centerpiece of the memorial was a white column, perhaps thirty feet tall, with a large bronze star perched on top. Fresh flowers had been placed at its base.

Layers of catastrophe had been overlaid on this landscape. During World War II—long before any nuclear reactors came along—the area around Chernobyl had been the scene of brutal fighting. As local partisans resisted the German occupation, the people suffered murderous Nazi reprisals, only to endure a horrific famine once the war was over.

In that context, it's hard to say that the accident in 1986 was even the worst thing offered up to Chernobyl by the twentieth century. Indeed, although the human dislocation caused by the accident was immense, its legacy in terms of illness and death is deeply ambiguous.

In the public consciousness, Chernobyl stands for cancer, deformity, and death. Even now, a quarter century later, there is no shortage of charities dedicated to the care of "Chernobyl children"—recently born kids suffering from cancer or birth defects attributed to the accident's aftereffects. But the Chernobyl Forum (a consortium including several branches of the UN and the governments of Ukraine, Belarus, and Russia) has argued that, after an epidemic of thyroid cancer among children living in the area during the accident, no measurable increase has yet been demonstrated in the region's cancer rates. The Forum's projection of excess cancer deaths in the future is surprisingly low, at about five thousand. Meanwhile, its estimate of the number of people killed by the accident's immediate effects stands at fewer than a hundred. Such estimates drive organizations like Greenpeace crazy, and they have produced their own numbers—of nearly a hundred thousand projected cancer fatalities, and sixty thousand already dead. Who knows, maybe the UN is the nuclear power industry's stooge.

More fundamentally, it's just hard to accept how little is known with any confidence about the disaster's effects, whether on people or animals. And it's hard to accept that the Chernobyl children may be the children of regular misfortune, not of nuclear fallout. That the accident's most traumatic effects may have been social and psychiatric, rather than radiological. That Chernobyl—and humankind's wretchedness—may not quite have lived up to our expectations.

\\\\\\\\\\\

Early the next morning, in the zone's only hotel, I awoke to the symptoms of acute radiation poisoning.

Inflammation and tenderness of exposed skin. Nausea and dehydration. Exhaustion and disorientation. Headache. Did I mention the nausea? I was still in my clothes, sprawled on top of a ruffled pink bedspread. The ceiling listed sideways in a sickening spiral.

I lay motionless, hoping for death, and stared upside down through the window above my head. Beyond the gauzy curtains, a massive Ukrainian dawn burst downward into the sky. It made me want to burst, too.

It wasn't radiation sickness. What I had was a bad hangover and a bit of sunburn. But I didn't see much difference.

I had found the nightlife in Chernobyl. Coming back from the war memorial, we had visited the outdoor "vehicle museum," a tidy grass parking lot with a fleet of military trucks and personnel carriers left over from the cleanup. Already slightly tipsy, we amused ourselves for a moment by dipping our radiation meters into the wheel well of an armored personnel carrier and listening to them scream, and then headed back to find the party.

The party was across the road from headquarters, in front of the hotel, and consisted of Dennis, Nikolai, and me, sitting on a bench in the parking lot. The hotel—it was more like a nice dormitory, really—was otherwise deserted. I'm sure you can still get good rates. I went up to my room and brought down some gifts: a Mets cap for Dennis, a pair of New York shot glasses for Nikolai, and a bottle of vodka for everybody.

We followed the strict custom that a bottle opened is a bottle that must be emptied—even though Nikolai wasn't drinking tonight and Dennis was too polite to outpace me. Toast upon toast seemed to improve my Ukrainian, and Nikolai's English, and the fluidity of Dennis's translation, and soon it was unclear to me which of us was speaking what. By this time it was completely dark, and my elbows had what I was certain were beta-radiation burns from leaning on the hood of the car, and we had somehow ended up in a bar.

There is a bar in Chernobyl, I thought. *There is a bar in Chernobyl.*

How we got there, or exactly where it is, was quickly lost in the fumes of my mind. I was deeply drunk. A lifetime lived in moderation had left me unprepared for this work. But if this was the price, I would pay it. I had found Chernobyl's only nightclub—even if it was little more than a bare, cinder block room with half a dozen people quietly slugging vodka and cognac out of tiny plastic cups.

"So, Dennis!" I shouted. "Is the zone a good place to meet girls?"

He nodded sagely. "There are many girls here," he said. "And they are all over fifty."

It's beyond cliché to suggest that drinking is the way to befriend Slavs, but it's also true. We left the bar at full stumble, arms slung around each other's shoulders, only partly because I couldn't walk. Nikolai, still sober, was proclaiming his enthusiasm for the project of pollution tourism. Most people came to Chernobyl just to get their two photographs, he said. They treat the staff like servants and leave. They never bother to find out what a nice place the zone can be.

I raised a nonexistent glass, and we came weaving into the parking lot, singing in Ukrainian at the top of our voices, exchanging a series of cavorting high-fives. I said goodnight to my brothers, and then somehow, in a single, fluid motion, fell up the stairs, down the corridor, through a locked door, and into bed. Which is where I found myself the next morning, feeling like a fishbowl brimming with bile.

At headquarters, breakfast was a reprise of the previous night's antipasto. I introduced a piece of cheese to my mouth, wet it with a teaspoon of water, and left it at that. Outside, I found Dennis and Nikolai. One look at my expression, and they both burst out laughing. Dennis shook my hand and smiled. "Next time, give me some more notice that you're coming," he said. "I'll show you the really good stuff. Maybe we can go in a helicopter."

The jerk. Surely I could stay on? Wasn't there still time for helicopters and canoes? But it was *not possible*. These things needed to be booked in advance. The permissions. An escort. And Dennis already had a group of

Ukrainian journalists out in the parking lot waiting to begin their visit. The nascent business of zone tourism carried on.

So learn from my mistakes. Plan on two nights.

In the car, I leaned gingerly against the seat, trying to disappear. Nikolai laughed again. There was still entertainment value in my hangover. He stamped on the gas, and we started for Kiev. It was another beautiful day for a drive. More glorious countryside, more checkpoints. Guards waving their excellently bulky Geiger counters over the car to test it for contamination. And detectors like phone booths, for us to hug, to test ourselves. And the road back to Kiev, through roadside villages, past pairs of men swinging scythes in the fields, and onto the highway, already swelling with the first weekend traffic streaming out of the city.

I wasn't done with the Exclusion Zone. In the back of my mind, a scheme was beginning to form. A scheme for a picnic near Strakholissya, the town I'd seen on the map. A scheme that would require Olena to help me borrow a rowboat. Maybe on Sunday?

But for the moment, the world was still half-spinning, and I couldn't look. I rolled the window down and felt relief stream in with the wind. Nikolai hugged the edge of the road as we picked up speed, and I leaned my head against the frame of the car and listened to the rising drone of the engine, eyes closed, mouth hanging open, gulping in the sweet, sunny air of the Exclusion Zone.

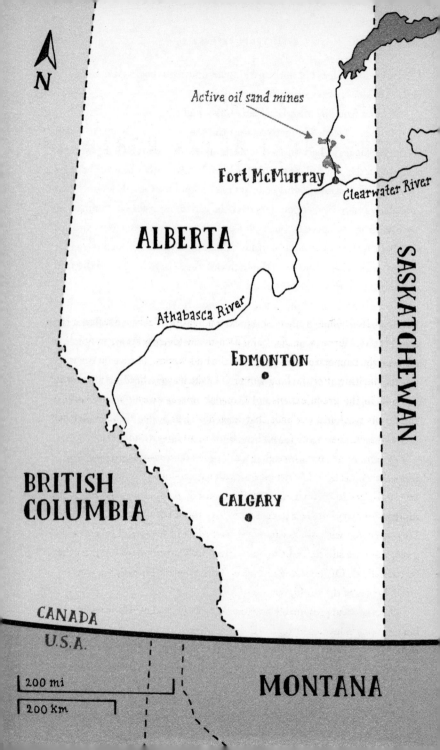

N

Active oil sand mines

Fort McMurray

Clearwater River

ALBERTA

SASKATCHEWAN

Athabasca River

EDMONTON

BRITISH
COLUMBIA

CALGARY

CANADA
U.S.A.

200 mi

200 Km

MONTANA

THE GREAT BLACK NORTH

On April 28, 2008, a group of some sixteen hundred ducks landed on a lake near Fort McMurray, Canada. It was a warm day for early spring in northern Alberta, the temperature reaching into the mid-sixties. The ice on the water was still melting after the long winter. The ducks were heading for nesting grounds in the green expanse of Canada's boreal forest—a vast band of coniferous trees and wetlands that stretches clear across Canada and that provides a summer home for half the birds in all North America.

Around these parts, though, a duck can't safely assume that a lake is in fact a lake. This lake, for instance, was actually a huge tailings pond owned by the Syncrude corporation—"tailings pond" being a term of art in the mining industry for "waste reservoir." As the birds touched down, they became coated with oily bitumen residue. Most of them sank. Others languished on the surface, waiting to be saved by human beings or videotaped by journalists. Of the sixteen hundred birds, fewer than half a dozen survived. Ducks of the world, beware of Alberta.

Syncrude had presumably hoped to keep its little duck holocaust private, but an anonymous tipster reported the incident and, before the day was out, the company had a full-blown public relations disaster on its hands. "Hundreds of Ducks Dead or Dying after Landing on Syncrude Tailings Pond," reported the *Western Star*, while the *Spectator* ran the cheeky "Tar Pond Dooms Ducks to Death." Within days, the scandal grew from mere corporate misfortune—"Syncrude in Hot Water over Duck Disaster" (*Windsor Star*)—to provincial government headache—"Duck Disaster Sinks Alberta Government's Credibility" (*Calgary Herald*)—to a matter of national import that demanded the prime minister's attention—"Harper Promises to Investigate Dead Ducks in Northern Alberta" (*CBC*).

This, then, is Canada—perhaps the only country where ducks have national, even geopolitical, significance.

But this isn't because the Canadian character is somehow uniquely sensitive to the welfare of its waterfowl. It's because the sixteen hundred—long may their memory live—had, with their deaths, scratched a festering sore on the Canadian national psyche. They had landed—and died—in something larger than a lake. Larger than a tailings pond. They had hit a grim bull's-eye in the world's largest and most controversial energy project, in the Middle East of the Great White North, in the cauldron of our energy future. They had landed in oil sands country.

\\\\\\\\\\\

Canada lives in the imagination of the United States as a benign, continent-size footnote, the brunt of conservative jokes about invasion and annexation, and the object of liberal daydreams about socialized medicine and sensible bank regulation. If there is an overarching consensus among Americans about their cousins to the north, it is that they are *like* Americans but nicer, probably smarter, and more loving of hockey.

Less well known is that Canada is a towering, earth-shaking, CO_2-belching petroleum giant. Let us keep our stereotype that Canadians are mild-mannered, but in terms of oil there is nothing moderate about them. They have it. With something like 175 billion barrels' worth hidden under the ground up there, the country is second in the world only to Saudi Arabia in proven petroleum reserves. The United States' number-one single provider of foreign oil isn't someplace in the Middle East. It's Canada.

A secret joy must surge through the heart of the US economy at this fact. Here on our very doorstep is a Persian Gulf without the Persians. A Saudi Arabia without the Saudis—or the Arabians. And Canada literally advertises this fact. In 2010, the Alberta government bought time on the huge screens of Manhattan's Times Square. "A good neighbour lends you a cup of sugar," one ad read. "A great neighbour supplies you with 1.4 million barrels of oil per day." It's enough to make modest, climate-change-fearing Democrats want to build pipelines.

Those 175 million barrels, though, come with a 170-billion catch. Most of Canada's oil—half of what it produces today and 97 percent of what it expects to produce in the future—isn't in the form of liquid petroleum, ready to be pumped out. It's oil *sand*, a thick, grimy sludge buried underground. And it takes more than sticking a straw in the ground to drink this particular kind of milkshake. It takes the world's largest shovels, digging vast canyons out of what was once Alberta's primeval forest; and the world's largest trucks, delivering huge quantities of the sticky, black sand into massive separators that need insane amounts of heat and water to boil the sand until the oil floats out of it, leaving behind—not incidentally, if you're a duck—unfathomable quantities of poisonous wastewater, which are then stored in tailings ponds of unusual size.

Got it? Environmentalists call it *dirty oil*, as if the stuff that comes out of the ground in Kuwait were somehow clean. But oil sands oil isn't dirty just because it requires strip-mining on a terrifying scale, or because it generates entire lakes of waste. It's also energy-intensive: you have to spend a lot of energy to separate and process the oil, much more than if you were simply pumping petroleum out of a well. So if you're passionate about carbon dioxide emissions and climate change—passionate about avoiding them, that is—oil from oil sands should give you the creeps. When you burn it, you're also burning all the energy that was used to produce it. The technical term is *double whammy*.

Engineers in the audience may argue that in terms of CO_2 emissions, oil sands are at worst a 1.25 whammy, depending on how you run the numbers. Nevertheless, a movement has coalesced around the goal of stopping oil sands development, with environmentalists determined to make Canada stop digging new Grand Canyons in its backyard. Leave the sticky stuff in the ground, they say, reasoning that, with the world already suffering for our overuse of fossil fuels, this is no time to be developing a new source.

But it's hard to hear that argument over the incredible grumbling sound coming from the collective stomach of the United States. It sees Canada's oil as a possible route to so-called energy independence, which is another way of saying "oil without Muslims," and it wants nothing more than for Canada to rip the green, boreal top right off the entire province of Alberta and shake all that black, sandy goodness directly into a refinery. And that drives environmentalists batshit crazy with rage.

\\\\\\\\\\\\

Fort McMurray lies in a splendid isolation of forest and swamp, nearly three hundred miles north of Edmonton, the provincial capital and nearest major city. As with most boomtowns, it's tempting to call Fort McMurray a shit-hole, but its attempt at wretchedness is halfhearted. For every corner of town that is dingy or low-rent, there is one that is tidy and clean. For example, Franklin Avenue: there is the Oil Sands Hotel, its yellow sign illustrated with large, orange oil drops. A narrow marquee boasts, CHEQUES CASHED, LOW RATES, RENOVATED ROOMS 99.00, ATM IN LOBBY, EXOTIC DANCERS MONDAY–SATURDAY 430–1AM. Across the way, as counterbalance, are the city hall and provincial buildings, a pair of sleek brick cubes that project an orderly municipal competence. At seven and nine stories, they are the tallest things in town. The next block down you'll find the Boomtown Casino, busy even at midnight on a Tuesday, as the people of Fort McMurray feed their oil sands money into slot machines.

Downtown sits on the triangle of land where the Athabasca and Clearwater Rivers converge and run north. But Fort McMurray is growing. Just across the Athabasca, a loop of fresh suburbs three times the size of downtown sprawls up the hill. In the eight years preceding the global economic slowdown of 2008, the city's population nearly doubled, to about a hundred thousand people. Housing is therefore exceedingly tight in Fort McMurray, and prices are closer to what you might expect in Toronto than in some town a five-hour drive from anywhere. Places to live are in such short supply, and the population drawn by oil sands work so transient, that some twenty-five thousand people—nearly a quarter of all residents—live in work camps provided by the oil sands companies. Which is to say, they don't really live here at all.

I arrived on a broad summer day, the sky smooth and bright and warm. I was staying with Don and Amy, an affable couple I had contacted through friends. Along with a teenage son, they lived in a two-story house in one of the recently built suburbs. Don was tall and thoughtful and wore socks with his shorts. Amy was small, dark-haired, and sprightly in a way that made her seem much younger than she was. They were in the full flower of middle age, spending their free time hiking and bicycling when the seasons allowed it. Hospitality seemed to come to them as a natural side effect of owning a house, and although they had no idea who I was or why I was there, they gave me my own bedroom upstairs and let me have the run of their fridge.

They both worked for oil sands companies: Amy for Suncor, Don for Syncrude. These are Canada's two primary oil sands companies, and each reliably pulls in billions of dollars in annual profits. Amy did leadership training, while Don was an engineer.

What, they wondered, was I doing in Fort McMurray?

I didn't want to say I had come to their town to see how the very two companies they worked for were ruining the world. It's this phobia I have about not seeming like a total asshole. So I gave them the long, squirmy version, something about environment and industry and seeing for myself and—

"Well," said Amy brightly. "We both work for the dark side."

The dark side?

Don scratched his head. "I don't know if you heard about our duck episode."

\\\\\\\\\\\

The rivers and forests that cradle Fort McMurray offer plenty of invigorating outdoor activities to visitors looking for that sort of thing. By the looks of it, you could do some great hiking or buzz the river on a Jet Ski, and I'm sure there's moose around that you could shoot. But the pollution tourist goes to Fort McMurray only for the mines.

It was a homecoming of sorts. I was born in Alberta (in Calgary), and although I left before I was two years old, it had always lingered in my imagination as that magical place—*the place I'm from*. This was my first time back in the province, and I intended to celebrate by seeing some torn-up planet.

I will admit to a certain excitement about it all, even though the responsible attitude, as a sensitive, eco-friendly liberal, would have been one of grave concern, or even horror. But I'm also the son and grandson of engineers: intelligent, bullshit-allergic men out of Alaska and South Dakota, men who lived by their knowledge of roads and of pipelines, and of rocks, and of how things get done. And though I inherited barely a trace of their common sense, I honor them how I can. How else to explain my almost sentimental enthusiasm for heavy infrastructure and industrial machines?

You could say, then, that I came to Fort McMurray with conflicted feelings about the oil sands, unsure of just how much filial gusto and faux-local pride were appropriate at the scene of a so-called climate crime. But this could be said about Canada in general. I was merely a walking example of the country's love-hate relationship with its own resources. The modest northern country where Greenpeace was founded had been declared an "emerging energy superpower" by its own prime minister, and in a spasm of vehement ambivalence, Canada was both pioneering the era of dirty oil and leading the fight to stop it.

\\\\\\\\\\\

Suncor's and Syncrude's main operations are located a quick jaunt up Highway 63, which runs parallel to the Athabasca, past hummocks of evergreen. About twenty-five miles out of town, the air starts smelling like tar. Suncor's business is hidden from the road, but Syncrude shows a little leg. As you get close, the trees disappear, and you pass a long sandy berm; one of Syncrude's flagship tailings ponds sits on the other side, a shallow lake of glassy wastewater.

I rolled down the window to let in the breeze, tarry and warm. The cracking thuds of cannon fire punctuated the air. It was the bird-deterrent system, the one that Syncrude had been a little slow to deploy in the spring of the previous year.

Let us hope that ducks find these noises either helpful or terrifying. Personally, I found it hard to tell where they were coming from. Had I been a duck, I would have wanted to land, to get my bearings and figure out just what the hell was going on. This also would have afforded me a closer look at the other bird-deterrent: a sparse posse of small, flag-like scarecrows that decorated the shore. Several more of the ragged little figures floated on lonely buoys in the middle of the lake.

The mines themselves were nowhere visible, but at the north end of the lake rose the Syncrude upgrading plant, the flame-belching doppelgänger of Disney's Enchanted Kingdom, built of steel towers and twisting pipes, crested with gas flares and plumes of steam. A hot, wavering stain of transparent yellow rose from one smokestack, drawing a narrow stripe across the sky.

Oil sands contain a heavy form of petroleum called bitumen, which must go through several stages of upgrading at a plant like this before it can enter a refinery. But before it can even be upgraded, it must be separated from the vast quantities of sand that are its host. This first step takes place mine-side, where the sand is mixed with water and then heated, separating out the layer of bitumen that clings to each grain of sand. You have here two issues: the use of massive amounts of water—in this case drawn from the Athabasca River—and the incredible volumes of natural gas required to heat it.

The separated bitumen is then piped to the upgrading plant, where—using yet another unimaginable amount of energy—it is put through a series of distillations and *cracking* processes to break it down into smaller, more manageable hydrocarbons. Only then can the result—called *synthetic crude oil*—be sent off to a refinery for the production of gasoline, jet fuel, and ziplock bags.

I hung a left, following the loop that would take me past the front gate, around the tailings pond, and back toward town. Just west of the plant was the sulfur storage area, though to call it a "sulfur storage area" is like calling the pyramids a "stone storage area."

One byproduct of Syncrude's industrial process is a monumental quantity of sulfur, for which it has neither a use nor a market. So it stores the stuff, pouring it into solid yellow slabs, one hulking yellow level on top of the last, building what is now a trio of vast, flat-topped ziggurats fifty or sixty feet tall and up to a quarter mile wide. Like everything else around here, they may be some of the largest man-made objects in history—but I had never heard of them before. A pyramid of sulfur just isn't news, I guess. They are less scandalous than a city-size hole in the ground, and only a very determined duck could get itself killed by one.

One day, though, Syncrude or its successors will see these vast—huge, monumental, gargantuan, monolithic—objects for the opportunity they are. Tourists of the future will summit their grand steps, and stay in sulfur hotels carved out of their depths, and sip yellow cocktails, and attend championship tennis matches at the Syncrude Open, for which the players will use blue tennis balls, for visibility on the sulfur courts. Thousands of years later, explorers bushwhacking through the jungles of northern Cameximeriga will stumble onto them and be dazzled by the simplicity of our temple architecture, at once brutal and grand, and will speculate about what drove us to worship sulfur above all other elements, and will see that the pharaohs were nitwits.

\\\\\\\\\\\

Although the mines sit at a breezy remove, their presence is felt everywhere in Fort McMurray. The economy and community thrum in tune with the ceaseless project of ground-eating. As you meander the streets, you begin to feel that you are an iron filing oriented along the field lines emanating from an immense subterranean magnet, and that everything and everyone in town is pointed toward it: the new bridge over the Athabasca, built to withstand the load of heavy equipment being transported to the work site; traffic lights that can be swung sideways out of the roadway to let oversize loads pass unhindered; the local high school (mascot: the Miners; motto: "Miner Pride"); the old excavating machine sitting on the lawn of Heritage Park.

You feel it standing on a wooded bluff overlooking the river, where the air stinks of bitumen oozing naturally out of the hillside, and where nearly a century ago the first hopeful entrepreneur tried to boil money out of oil sands. And you feel it downtown at the Tim Hortons, where white pickup trucks line up around the corner to get their coffee and doughnuts. Each white pickup truck carries someone on his way to work at the mines, and each white pickup truck has a tall, whiplike antenna sprouting from its bed, and they are not antennas but safety flags. Without one, even a large pickup truck may go unnoticed by the behemoth sand haulers in the mine, and be crushed.

Even at leisure, people in Fort McMurray live out an echo of their industry, taking their minds off the noisy machines of the mines by churning through the countryside on other noisy machines, like all-terrain-vehicles and snowmobiles (known as *sleds*).

"Ninety percent of people who live here have at least one ATV or sled," said Colleen, the young woman behind the counter at the off-roading store.

She and her colleague Adam were Fort McMurray natives, rarities in a city overrun by outsiders coming for work, and they had a blasé defensiveness about their hometown. Colleen seemed almost to rue the economic boom that had transformed it. "The recession sucks and all, but in ways it's amazing," she said. "Now you can go to a restaurant and not wait three hours. You can get a doctor's appointment. Before, if your car broke down, it would take nine weeks to get it fixed. The quality of life was getting really low before the recession happened. Everything was a struggle."

But that didn't mean they thought the oil sands themselves were a bad thing. "Fort McMurray is what's powering all of Canada, and we don't get the recognition," Colleen said, picking up a tiny brown dog bouncing at her feet. "I think that whole 'dirty oil' thing comes from a lobbying group in Saudi."

"The ducks," Adam said, completing the conspiracy theory.

Colleen snorted. "Yeah, fuck! There's so many more important things. Like consumer waste!"

\\\\\\\\\\

Through Fort McMurray Tourism, anyone who signs up a day ahead and forks over forty bucks can take an oil sands bus tour. *Oil sands bus tour*—are there any four words more beautiful in the English language? Someone was finally seeing the light on this pollution tourism thing. I signed up.

The bus tour didn't leave until the following morning, so I had a lonely afternoon to kill. I called my girlfriend. The Doctor. She always knows what to do in these situations. She has a peculiar kind of common sense that includes the possibility that spending your days roaming oil sands mines and nuclear disaster sites might be a good idea.

"Remember," she said over the phone, "you're supposed to be on vacation."

Right! I was a tourist. And although the world's industrial eyesores and ecological calamities generally languish unattended by gift shops and welcome centers, Fort McMurray is a forward-thinking town in this regard. I made for the Oil Sands Discovery Centre, a family-friendly museum for those interested in the local industry.

The OSDC represents some of the best industrial propaganda in the world. (Which I mean as a compliment. *You* try writing the brochure for Mordor.) Its gift shop is a gift shop among gift shops, an emporium thick with toy giant dump trucks, kid-size hard hats, watercolor prints of gigantic machines, and truck-themed socks. I grabbed an armful of goodies. At the register, I made the find of the day in a bin of impulse buys: a tiny, plush oil drop with yellow feet and googly eyes. Who knew petroleum could be so adorable?

Into the exhibits, where I spent the next several hours in a state of fizzing excitement over scale models of dragline shovels and bucket-wheel extractors, over containers holding liquid bitumen in different states—room temperature, heated, diluted—with rods to stir the stuff and feel the different viscosities. Not to mention the 150-ton oil sands truck parked inside the exhibit hall. I climbed two stories up, into its cab, and sat in the driver's seat, wrenching the steering wheel back and forth.

And now let us praise the Dig and Sniff, in which a small mound of raw oil sand is displayed under a plastic dome. The Dig and Sniff invites you simply to *dig*, using the rod built into the display—and then, having dug, to *sniff*, through the small opening in the dome. *Dig and Sniff!* With a name of such economy and force, it commands you to action, granting you a direct experience—modestly tactile, safely olfactory—of the oil sands themselves.

A young boy worked the scraper. "This thing is cool!" he cried, sticking his nose into the dome. "Dad, come smell the oil sand! The Discover Center's *fun*." We were living inside a commercial for the OSDC. I took my turn at the stand, ready to get down to business.

I dug. I sniffed.

Frankly, it didn't smell like much. Maybe it needed a fresh batch of sand. But had I not already learned something? That oil sand may sometimes lose its aroma?

You could be forgiven for assuming—it would be weird if you didn't—that the OSDC was created by the oil sands companies themselves, as a temple to their own name. But among its many triumphs in industrial propaganda, surely the greatest is that it is actually a government facility, operated and administered by the province of Alberta itself. You can draw your own conclusions about what this seamless collaboration says about the relationship between oil and government around these parts.

Underneath all the excitement, though, there was a sour note—a defensive, self-conscious tone that sometimes crept into the wall copy. I could feel the exhibit designers grudgingly trying to account for that one spoilsport in each group, the one who would be asking over and over about the trees, and the rivers, and the ducks.

Toward the end of the galleries, past a backwater of displays about environmental responsibility and the future of clean energy and other boring crap, I found the Play Lab, a colorful area partially screened off from the rest of the hall by a metal space-frame. Child-size tables and chairs sat in the center of the room, attended by a wardrobe of hard hats and jumpsuits available on loan to the tiny oil sands engineers of tomorrow.

Ignoring the cues that I fell somewhat outside the Play Lab's target demographic, I charged in, blazing my way through the *PUMP IT* exhibit—a wall of clear plastic pipes with valves to twist and a crank to turn—before settling in for a spell at *DIG IT*, which featured a pair of toy backhoe shovels and a trough filled with fake oil sand.

Neeeat!

The last section of the Play Lab was *GUESS IT*, a large grid of spinning panels printed with questions on one side and answers on the other. Somewhere an exhibit designer, worried about how much fun the rest of the Play Lab was, had caved in to the didactic urge. I read the first panel.

Bitumen is a very simple molecule. True or false?

Duh! We're talking about hydrocarbons, here. False. Next question.

Oil sand is like the filling in a sandwich. True or false?

Uh, true?

True. The top slice is overburden, oil sand is the gooey filling, and the bottom slice is limestone. Yummy!

I no longer had the lab to myself. An elderly couple had entered and, after a cursory look at the shovels, were now having a go at *PUMP IT*. I turned back to *GUESS IT*.

Who is responsible for protecting the environment? a) the government, b) the oil sands companies, c) everyone.

It was that defensive tone. I didn't need to turn the panel to know what *GUESS IT* wanted me to say. The only question was whether children were really the Play Lab's target audience after all.

\\\\\\\\\\\

The highway north of Fort McMurray is so small, relative to the thousands of workers who need to get to the work sites every day, that traffic can be terrible, especially during shift changes. So the oil sands companies hire buses to ferry workers to and from town. Ubiquitous red and white Diversified Transportation coaches ply the highway in pods. That an industry partly responsible for Canada blowing its emission-reduction goals has a thriving rideshare program is just one of the tidy, spring-loaded ironies that jump out at you here.

The Suncor bus tour leaves from in front of the OSDC—I stole in for a quick taste of the Dig and Sniff—and it employs one of those same Diversified buses, re-tasked for our touristic needs. Mindy, our perky young tour guide, popped up in front and asked us to buckle our seat belts. "Safety," she said, "is one of our number-one priorities." The driver gunned the engine and we were off, about to be taken, Universal Studios–style, through an open wound on the world's single largest deposit of petroleum. What soaring cliffs and hulking machinery did the day hold for us?

The bus was nearly full, mostly with families and seniors—people who looked like they had seen the inside of a few tour buses. A quartet of old ladies giggled like they were on a Saturday-night joyride. Sitting next to me was a Mr. Ganapathi, an old Indian man with a single, twisting tooth jutting from his lower jaw.

"You are married?" he asked.

I wasn't, I said. But I thought of the Doctor. It wasn't a bad idea.

By now we were passing along the eastern edge of the large tailings pond in front of Syncrude.

"Is this where all those ducks got killed?" a man asked his wife.

"Oh, we've had more fuss over those ducks!" she said.

There had indeed been more fuss. The governments of Canada and Alberta had decided to prosecute Syncrude for failing to repulse the ducks from the tailings pond. There would be a not-guilty plea, and complaints from Syncrude that it was being unfairly prosecuted for what amounted to a mistake but not a crime, and counter-complaints from environmentalists that Syncrude was getting off easy. In the end, Syncrude would be found guilty and fined $3 million—$1,868 Canadian for each duck. And if those sound like expensive ducks, keep in mind that in 2009 Syncrude made $3 million in profit every single day.

We stepped down from the bus near the Syncrude plant—it hissed in the distance—to visit a pair of retired mining machines. You needn't take the bus tour to see them, though, as they are probably visible from space. I had never seen such machines. A dragline excavator stood on the right; on the left, a bucket-wheel reclaimer.

These days, oil sands mining uses shovels and trucks in a setup that has a nice scoop-and-haul simplicity to it. But this system is relatively recent. Previously, companies used a system of draglines, bucket wheels, and conveyor belts. With a dragline excavator (a machine probably bigger than your house), a bucket-like shovel hanging on cables from a soaring steel boom would gather up a bucketful of sand—and we're talking about a bucketful the size of . . . the size of . . . hell, I don't know. What's bigger than an Escalade but smaller than a bungalow? *Big*, okay? The dragline would swing around, using the huge reach of the boom, and drop the sand behind it. It would then inch along the face of the mine, walking—actually *walking*—on gigantic, skid-like feet, repeating the process over and over, leaving behind it a line of excavated sand called a *windrow*.

Then the reclaimer would come in, turning its bucket wheel through the sand in the windrow, lifting it onto a conveyor belt on its back, which fed another conveyor belt, and another, transporting the sand great distances out of the pit. There were once thirty kilometers' worth of conveyor belts operating in Syncrude's mine, and if you've ever tried to keep a conveyor belt running during a harsh northern winter—who hasn't?—you've got an idea of why they finally opted for the shovel-and-truck method.

To approach the bucket-wheel reclaimer was to slide into a gravity well of disbelief. It was difficult even to understand its shape. It was longer than a football field, battleship gray, its conveyor belt spine running aft on a bridge large enough to carry traffic. The machine's shoulders were an irregular metal building several stories tall, overgrown with struts and gangways and ductwork, hunched over a colossal set of tank treads. A vast, counterweighted trunk soared over it all, thrusting forward a fat tunnel of trusses that finally blossomed into the great steel sun of the bucket wheel.

The wheel itself was more than forty feet tall, with two dozen steel mouths gaping from its rim, each worthy of a tyrannosaur, with teeth as large as human forearms. I stared up at it, nursing a euphoric terror, imagining how it once churned through the earth, lifting ton after ton of oily sand as it went. There was something wonderful about the fearsome improbability of the reclaimer's existence. It was the bastard offspring of the Eiffel Tower and the Queensboro Bridge, abandoned by its parents, raised by feral tanks.

As my tourmates took pictures of one another standing in front of the behemoth, I walked back to the bus, where the driver was standing with his hands in his pockets. His name was Mohammed. The Suncor bus tour was only a minor part of his job. He spent most of his days ferrying workers to and from the mines. When I asked why he didn't choose to drive one of the big trucks instead of a bus, he told me he wasn't interested.

"But you could make a lot of money," I said. The salary for driving a heavy hauler started at about a hundred thousand dollars—more if you worked a shovel.

He smiled. "The pollution. Especially at the live sites, Suncor and Syncrude." He thought the air coming off the upgrading plants was bad for your health.

"But you breathe that air anyway," I pointed out. "You drive onto those sites all the time!"

He laughed. "Yeah!"

The supposed centerpiece of the Suncor bus tour is of course Suncor itself. We entered from the highway, the air sweet with tar, and drove toward the Athabasca River into an area invisible from the road. My oil sands fever was reaching its crisis. The upgrading plant slid into view, a forest of pipes and towers similar to the Syncrude plant, but nestled next to the river in a shallow, wooded valley.

It was getting hard to pay proper attention to the scenery. Mindy had been keeping up an unrelenting stream of patter, a barrage of factoids that, despite its volume, managed to be completely uninformative. I found it difficult to follow her, even with my inborn enthusiasm for pipes and conveyor belts and giant cauldrons of boiling oil.

The green building houses the fart matrix. It uses 1.21 gigawatts of electricity every femtosecond.

The what matrix? Wait, which tower was—

. . . three identical towers of different sizes on the far side of the plant—can everybody see?

No, wait, which?

Good. Those are where the natural solids ascend and descend twenty-one times per cycle, each cycle producing ten metric tons of nougat, which is sold to China, because it can't be stored so close to the river. The interiors of the towers have to be cleaned every two weeks using high-pressure ejaculators. Wow!

As we passed over the river—the river from which Suncor extracts about 180 million gallons of water per week—Mindy threw us a few bones of actual information. One point five million barrels of bitumen come out of the oil sands every day, she said, and Suncor had four thousand employees working on the project, which ran twenty-four hours a day, 365 days a year.

Underneath the avalanche of information, we were becoming dissatisfied. When would the drive-by of the upgrading plant and the mine's logistical centers end and the actual oil sands tour start?

"Are we going to get close to one of these trucks?" growled a man in the back.

Mindy smiled. "I'm going to try!" she said. But of what her trying consisted, we will never know.

The bus continued down the road, past a few nice pools of sludge, the occasional electric shovel dabbling in the muck, and a couple of flares. In a bid to drown our curiosity before we mutinied, Mindy had begun a spree of pre-emptive greenwashing. Suncor was required by law, she told us, to "reclaim" all the land it used, meaning it was supposed to restore it, magically, to its state before the top two hundred feet of soil was stripped off and the underlying oil sands pulled out. As for the Athabasca River, if we were worrying about whatever it was that everyone was worrying about, we shouldn't.

"We're very limited in terms of what we can take during times of low flow in the river," she said.

Thank goodness. And had we noticed all the trees? Suncor had already planted three and a half million trees, she chirped. There were Canadian toads, *Bufo hemiophrys*, living fulfilling lives on this very land.

We had reached the far outside edge of the mine—a dark rampart of earth. A huge chute was built into the embankment—it was the hopper that fed the oil sands into the crusher. It sat distant and lonely, unvisited. Mindy checked her boxes as we passed: hopper, crusher, building, pipe, and we left it behind. The bus parked and we were allowed to descend, for the inspection of a large tire sitting in the parking lot of the mine's logistical headquarters.

We weren't going to get the merest peek into the mine. Here on the oil sands bus tour, we weren't going to see any trucks in action, any shovels, any actual *oil sands*. Here I was, ready to embrace some corporate PR with open arms, and even I thought it sucked.

The air reeked of tar. I had a headache. We got back on the bus. Mindy had some more information for us, something about how every ton of oil sand saves a puppy. She did not seem to have any realistic enthusiasm for oil sands mining, only a plastic version of the touchy, defensive pride endemic to the entire venture of oil sands PR. It's just distasteful to watch an oil company try to prove that it is not only environmentally friendly but also somehow actually in the environmental business. Instead of straight talk from a man with a pipe wrench, we have to tolerate oil company logos that look like sunflowers, and websites invaded by butterflies and ivy. (As of this writing, www.suncor.com presents the image of an evergreen sapling bursting through a lush tangle of grass.) Who are they trying to convince? Themselves?

On the way back to the upgrading plant, I noticed some activity next to the hopper, on the high rampart above the extraction facility. There were a pair of haulers backing toward the chute, each piled high with oil sand.

I clambered over Sri Ganapathi, straining for a clear view through the far side of the bus, snapping pictures as one of the dump trucks began to raise its bed to drop its cargo into the chute. But as it did, we passed behind a building and the scene disappeared. Mindy was going for the green jugular, telling us how Suncor had planted so much vegetation on its land that deer came to live there.

"There's no hunting allowed," she said. "So they're pretty happy." Suncor, you see, is not a multibillion-dollar petroleum company, but a haven in which deer and toads can live in peace. I wanted to spit.

The view came clear and I saw the second truck. Four hundred tons of sticky, black earth—a solid mass as large as a two-story building, and enough to make two hundred barrels of oil—slid smoothly off its upturned bed and down the maw of the hopper. I had the sensation of having seen an actual physical organ of the animal otherwise known as *our voracious appetite for fossil fuels.* The appetite belongs to a body—a body with many mouths, some of them built into the sides of open pits in Alberta.

The trucks lowered their beds, heading out for the next load, and the next. I had seen the human race take a tiny bite out of the world. The bus drove on. Nobody was watching.

\\\\\\\\\\\

"So, are we raping the planet?" asked Don.

We were sitting in the living room.

Based on the morning's utter bust of an oil sands bus tour, I said it was hard to declare with any certainty whether he and Amy were in fact raping the planet. I did hint, though, that there was room for competition in the oil sands bus tour niche.

After so much mealymouthed blather on the tour and at the OSDC, it was refreshing to talk to Don. But even he seemed fundamentally ambivalent. Don was an oil sands engineer, but he also had a degree in environmental science. He had begun his career on the reclamation side, and he talked eagerly about what was possible with a former mine site—even if his own company had only begun to reclaim the areas it had dug up.

"You can put overburden back in the mine at the end," he said. *Overburden* is the word used to describe the earth that is stripped off to reveal the resources underneath. (It's tempting to draw conclusions based on this word—that strip-miners see the landscape and forests only as "burdens.") In the reclamation process, the overburden, now free of vegetation, can be tossed back in the hole to help patch it up.

"Then you do replanting," Don continued. "Get the hill made, get it sculpted, build little lakes and marshes." He described the sequence of plantings that would follow, slowly restoring the land to something like what had been there before. And just like that, as if icing a cake, you could have your environment back.

But Don said he was better as a geologist than as an environmental scientist. So now his job was to build Syncrude's geological model, based on test data from areas to be mined in the coming years and decades. Bitumen richness, water content, grain size, rock types—there were dozens of measurements. Don integrated it all into a database that would allow the company to decide exactly where to mine, where to set its pits and its benches, where to put the shovels.

"I'm in awe of that," he said. He was in charge of the mining database of one of Canada's most profitable companies.

But there was an undercurrent to his enthusiasm. "I'm part of the mining process instead of part of the solution to fix it up afterwards," he said. "The budget for reclamation is so small compared with the profits they make." He shook his head. "They should be dishing out more." And indeed, only a microscopic portion of oil sands land has ever been certified by the government as reclaimed.

The answer, he thought, was stronger environmental regulation. But the Alberta government would never make it happen.

"They're getting zillions of dollars of royalties," he said of the province. "If you've got land, the government of Alberta will let you go in and take the oil out. They're interested in profits."

The late northern dusk had finally descended. The living room was getting dark.

"Do *you* think you're raping the planet?" I asked.

Don exhaled. "In terms of pollution, no, we're not," he said. "There's people downstream who say they're getting cancer from the oil sands operations, but we're not even putting anything in the water." But although he didn't buy claims of carcinogens in the Athabasca River, Don was no climate change skeptic. A huge amount of fuel was being burned to mine oil sands, and to extract and upgrade the bitumen—which meant a huge amount of carbon emissions. And those carbon emissions worried him.

"I once saw a map of CO_2 emissions in North America," he said. "There was a big fuzz up around Fort McMurray. The CO_2 from Fort McMurray is probably the same as from all of Los Angeles."

It seemed impossible. Could Fort McMurray really have carbon emissions similar to those of a city literally a hundred times its size?

Don had a way of saying things I might expect from an environmental activist—yet he was a man who spent his days helping the pit get wider. He embodied, far more than I did, Canada's contradictory feelings toward the oil sands and the consequences of their extraction.

But we all share in the paradox. Anyone does who both takes part in civilization and cares about the environment. Civilization sustains and protects us as individuals and communities, but it is more than a mere system for shelter and sustenance and order. It is what we are. The unit of the human organism is not the individual but the society. For better or worse, isolated individuals cannot sustain or further the human race. Only in society does it survive.

Today that society is an industrial one, resource-hungry and planet-spanning, growing so inefficiently large, we believe, that it is disrupting its own host. It is not strange, then, that some individuals of that society should question its integrity. They wonder whether the very thing that allows them to exist—the thing that they *are*—is not somehow rotten at its core.

This is the love-hate relationship in which we are all now engaged, and it is the basis for the entire spectrum of our individual decisions as they relate to the environment. Whether we're talking about recycling, or voting, or consumer choices, or political agitation, or radical efforts to live off the grid, these are all attempts to square the circle, to mitigate—or, more often, to atone for—our individual role in the disquietingly unsustainable system that keeps us alive. It's not just about living sustainably. It's about being able to live with ourselves.

As for Los Angeles, Don had his numbers wrong. Fort McMurray does not emit the same amount of carbon as LA. It emits twice as much.

\\\\\\\\\\

With the bus tour such a bust, I turned to finding a scenic overlook. I headed for Crane Lake, a Suncor reclamation site that seemed like a good starting point for some creative sneaking.

The word *reclamation* gets tossed around a lot in these parts, and not only in Don's living room. It is an important concept for anyone who doesn't want to feel too bad about strip-mining. Reclamation requirements use the vague guideline of "equivalent land capability," which means, according to the Alberta government, that reclaimed land has to be "able to support a range of activities similar to its previous use."

And that's the key here—*its previous use*. What, previously, was the use of an undisturbed boreal forest? What if its main use was to remain undisturbed?

I drove. I was in my little rental car, underneath a thick sunshine that was pushing back the afternoon's storm clouds. The highway was slick with rain and heavy with traffic. It was the beginning of the evening shift change. Work in the mines is divided into two shifts per day, and every twelve hours fresh battalions of truck drivers, shovel operators, plant workers, and engineers come hurtling up Highway 63. The road is long and straight, and the waves of pickup trucks and red and white buses had worked up to an insistent, humming speed. It was at that moment—as I approached the turnoff for Crane Lake, followed by a speeding phalanx of cars and buses—that I saw the ducks.

They came waddling onto the highway from the right shoulder, from the direction of Crane Lake. A mother and six ducklings.

A black sports car had just zipped past me and slotted back into the right lane. I was certain it was going to tear right through them, leaving them in twisted pieces; and that I, unable to stop, would mow through the survivors; and that if by a miracle there were still survivors after that, they would surely be obliterated by the wall of chartered coaches breathing down my neck. After so much talk of ducks and duck deterrents, of duck death and duck lawsuits, I was now about to help write the next chapter of Syncrude's environmental record, and that chapter was going to be written in blood, the blood of ducks, here on Highway 63, during the shift change.

It was over in seconds. The driver of the sports car braked and veered left, clearing the ducks by a few feet. Spooked, they turned and waddled back the other way, directly into my path. I found my moral sense neatly congruent, if only for a moment, with the needs of Syncrude PR. I swerved onto the shoulder, also missing the ducks, but spooking them again as I blew by and sending them back into the middle of the highway in disarray.

In the rearview mirror, I saw ducklings turning in every direction as their doom approached at seventy-five miles an hour in the form of a looming passenger bus—possibly driven by a man named Mohammed—riding abreast with a big white pickup truck and followed by more traffic behind. There was no leeway, no room for them to swerve. With horror, I imagined the bus careening into the ditch, rolling onto its side.

And then, somehow—it didn't happen.

The bus leaned forward, lumbering to its knee as it slowed. The pickup truck made a languid weave halfway out of its lane. And the rest of the oncoming column seized up and stopped. As the scene dwindled in my mirror, I saw the mechanized army of the Syncrude evening shift pause, like Godzilla offering Bambi a bouquet of daisies. And there they waited, patiently, as the ducks reformed their little rank and waddled off the highway back into the woods.

\\\\\\\\\\

Crane Lake is a nice spot, enclosed by a belt of young forest, with reeds clustering along its swampy shores and a nature trail running a mile circuit around the lake, through tall grass and wildflowers. The only footnote to the idyll is that the entire place stinks of oil sand, the same heady aroma that you would smell at a restaurant if the waiter set a bowl of bubbling tar on your table. The trick to experiencing Crane Lake, then, is to appreciate this smell as part of the environment, to remember that it's coming off of oil sand that God himself put in the ground—even if it's humankind that decided to rip it open and expose it to the air. As for the constant, popping reports of nearby bird-deterrent cannons, if they weren't enough to bother the birds that had come to take the waters at Crane Lake, then why should they bother me?

Forget that Crane Lake is called Crane Lake, though. It should be called Duck Lake—or maybe something punchier, like Suncor Ducktasia Lake. It is nothing less than Suncor's duck showcase. No nature area has ever been so completely tricked out with signs calling attention to what a lovely little nature area it is. There are duck blinds, and a duck-identification chart from an organization called Ducks Unlimited, and a good number of actual ducks present on the lake, possibly including several I had recently failed to murder.

So ducktastic was it that I began to wonder whether Suncor was trying to stick it to poor old Syncrude, with all its duck problems, just up the road. Surely some Suncor PR rep had hoped for a newspaper headline proclaiming, "Suncor, Neighbor to Duck-Destroyer Syncrude, Offers Clean Water, Reeds, at Waterfowl Haven."

I set out on my hike, keeping the lake on my right, ambling through a spray of purple wildflowers. There were dragonflies, again, and mosquitoes, too—snarling, clannish mosquitoes of the Albertan variety, with thick forearms and tribal tattoos. But I was ready. Don had lent me a bug jacket—a nylon shirt with a small tent for your head and face—and I had armed myself with enough spray-on DEET to poison a whole village. That is to say, I was happy, and ready to bypass all this man-made nature and find my scenic mine-overlook.

Making my way over a small wooden footbridge that spanned a swampy inlet, I was steered southward along the east shore of the lake by a thick forest of young trees on the left. A wooden bench, with grass growing up between the boards of its seat, faced the water. Silence reigned, except for the gentle rustle of the breeze and the constant sound of cannons. I had the place to myself.

But the farther I went down the path, the more the Crane Lake experience started to chafe. All this had been put here on purpose—*sculpted*, as Don had said. It was too neat. Too self-contained. Halfway down the east side of the lake, I turned to face the dense thicket of young trees that hemmed in the path. From a conspiracy-theory point of view, I reasoned, the very impenetrability of the forest here made it all the more likely that there was something interesting on the other side, perhaps something spectacular, or even hellish.

Ten seconds in, I had lost sight of the lake and the path, crashing through the trees, pushing branches out of the way, plowing through thick spiderwebs that collected on my face-tent. After a few more minutes of bush-whacking, I began to doubt that this was such a good idea. Everywhere I looked, the world looked the same: crowded stands of tall young trees closing in. I wasn't even sure which direction I had come from. I concentrated on the fantasy of breaking through the trees at the top of a magnificent cliff, looking out over the mine, trucks rumbling to and fro.

I saw light in the distance, through the trees, and went toward it, crossing a small clearing, then plunging back into thick overgrowth and more trees. I jumped a small ditch or stream, heading toward what seemed like a large, open area. It was close. I climbed a small rise of high ground, and it gave way like mud, my foot sinking down into it. I hopped forward, pulling my foot out, and saw sky ahead. Readying a mental fanfare, I broke through the tree line.

There was no vista. No overlook. No oil sands. Instead, I found myself standing on the edge of a cozy little wetland, swampy water winking in the sun.

Crap!

The way was utterly blocked by this revolting picture of nature in repose. I turned back in disgust. It was the sinister hand of Suncor at work, several moves ahead of me, drawing me in with the siren song of bird-deterrent cannons—and the drone of distant machinery, if I wasn't imagining it—only to throw wetlands in my path.

And now I was lost. Half-blind and overheating inside the face-tent, I walked in what I hoped was the direction of the lake, branches tearing at me. The mosquitoes circled, cracking their knuckles and waiting for that moment when the human, undone by panic and claustrophobia, tears off his bug jacket.

Finally, I saw the muddy rise I had sunk my foot into on the way over— a single landmark in a leafy wasteland—and staggered back toward it. About to cross over it again, I stopped short.

I could see my footprint from before, right in the center of the mound. It was swarming and alive. The small ridge was actually a great anthill. I bent over and looked into my footprint. Ants poured through it in chaos, frenetic in their attention to the fat, wriggling grubs, tumbling over them, picking them up, extricating them from the crater, the giant breach in their city wall. Sorry, guys.

\\\\\\\\\\

Crane Lake was pleasant in its way, but it was the merest green speck on a huge landscape of unreclaimed and active mine sites. Nor was it even a true test case. I later talked to Mike Hudema, of Greenpeace Canada, and he scoffed at the very notion of reclamation.

"When we destroy an area, we can't put it back," he told me over the phone. "We don't know how to do it. We can create something . . . but it's not what was there. It's not the same, and the way that life in the area reacts to it is also not the same."

That a guy from Greenpeace would be skeptical of mine reclamation was no surprise. More interesting was his contention that Crane Lake was never a mine site in the first place.

"It's basically reclaiming the area where they piled the dirt," Hudema said. "So it's not actually reclaiming a mine site. It's not reclaiming a tailings lake."

Hudema was that rare person who had been camping in the oil sands mines. One sunny autumn day, not long after my visit, Hudema and several of his colleagues had gone for a walk through Albian Sands, an oil sands mine owned by Shell.

Of course, no group of Greenpeace activists can go strolling through a mine without chaining themselves to something. In this case, they attached themselves to an excavator and a pair of sand haulers and rolled out a large banner reading TAR SANDS—CLIMATE CRIME. The entire mine was shut down for the better part of a shift, and Hudema and company spent thirty-some hours camping out on the machinery before agreeing to leave. (Later Greenpeace oil sands protesters met with arrests and prosecution.) The protesters' purpose—what other could there be—was to make the news, to raise awareness, to convince the world that there was something at stake worth getting arrested for. In them, Canada's love-hate relationship with the oil sands had most fully flowered into hatred.

But I also think of them as a breed of adventure travelers, and I thought Hudema might be able to share some tips for future visitors to Fort McMurray. Should hikers pack their bolt cutters?

"Well, unfortunately that's the part I can't talk about at all," he said. "It's sort of a general rule at Greenpeace that we never talk about how we get onto premises, because the question of why we go is much more important."

What a disappointment. I had expected pointers, even war stories. Weren't we colleagues of a sort? Didn't we share a profound fascination with the destroyed landscape of the oil sands mines—even though his fascination was politically engaged and mine was mainly witless?

Think, I thought. *Think of some question that will really capture his experience inside the mine.*

"What did you eat?"

"We brought all our own food in with us," he said, "and so we ate a variety of different things."

A variety of different things? It seemed like an evasion. I closed in for the kill.

"Does that mean sandwiches?" I asked.

"I don't really want to comment in terms of exactly what we ate," he said.

Although he refused to talk about access, or sandwiches, Hudema was willing to give me his impressions of the mine itself. "A barren moonscape," he said. "There is nothing but death. There's nothing living. All of the trees, all of the brush, everything above the earth's surface has simply been pushed away. All of the rivers have been diverted, all the wetlands completely drained. You just have these machines, larger than any on the planet, that just carve into the earth, three hundred and sixty-five days a year, twenty-four hours a day. And so from a visceral point of view, it's a horrific experience."

"Was there a sense in which you found it perversely beautiful?" I asked.

"Um, no," he said. "I would never use that word to describe it. It's just a place that is devoid of all life. A barren, barren moonscape. And you're constantly reminded of what used to be there. Or what should still be there."

What should still be there. That was the crux of it, I thought. The beauty or ugliness of a place didn't have that much to do with what it looked like. Even a moonscape could be beautiful—if it were on the moon. And who would deny the beauty of a desert, no matter how barren or harsh? Beauty depends on what we think is *right.* How else could we have come to think that unnatural objects like cities or farms or open roads were beautiful? That's what I wanted to see. The rind of beauty that must exist in every uncared-for corner of the world.

\\\\\\\\\\

Elevation. That's what you need. I hired a plane.

We took off straight into the sun, riding a little four-seat Cessna, and arced north, bringing downtown Fort McMurray under our right wing, and then its suburbs, newly carved out of the forest—Don and Amy's neighborhood. A clean boundary defined the edge of development, beyond which evergreen trees and muskeg swamp stretched out to the sky.

Terris was my pilot. Boyish and friendly, with broad, angular features and a strong Canadian accent, he had been in Fort McMurray for only a few months and earned his living by giving flying lessons and the occasional tour. During the boom of the previous decade, he had flown charters out of Edmonton. It had all been oil business, he told me, carrying executives and engineers up to private airstrips that the oil companies maintained on their lands. "The runways at Firebag and Albian are nicer than the Fort McMurray airport," he said. Engineers would come from as far away as Toronto and stay for a two-week shift before flying home to take a week or two off duty. It is a common cycle in Fort McMurray, except that most workers do it by car, driving back and forth to Edmonton along Highway 63.

Oil prices had fallen with the recession, though, and the oil sands business had entered another of its cyclical downturns. Terris's corporate work had dried up.

"So now I'm back in the bush," he said.

Fort McMurray dwindled behind us. The sun was low, behind a curtain of haze, the earth dusky. Sliding toward us were the sulfur pyramids of Syncrude, their full dimensions even more impressive from the air, a footprint five city blocks to a side.

"I have one flying student who's a Suncor engineer," came Terris's voice over the headset. "He was complaining about how people give the oil sand companies a hard time about polluting the Clearwater River. He said, 'The Clearwater River is one of the most *naturally polluted* rivers around.'" Terris was smiling. "The guy said, 'It's been leeching bitumen into the water for three million years. We're just doing the same thing!'"

We all have our ways of feeling like part of the natural order, I guess.

I could now see a low mountain of dry tailings that Don had told me to look out for, a huge heap of sandy mine waste that, like everything else around here, was one of the largest man-made objects in the world. It was so large that it was hard to tell where the tailings ended and the non-tailings landscape began. Beside it was a graphite-colored tailings pond, a mile and a half long, with a single boat floating motionless on its surface.

"People have really different reactions to seeing the mines," Terris said. "One group I had said it was the most horrible thing they had ever seen. And then you'll get engineers up here, and they just say it looks like a mine."

As we considered circling back for another look, the radio crackled to life.

Private aircraft, maintain minimum distance and altitude from Syncrude plant operation.

It was Syncrude security. The company had its own aircraft control. Terris grimaced. "I was hoping nobody would be home." But it didn't matter. Already we could see Suncor.

It loomed in the distance. Rather, it did the thing that is like looming but is actually its opposite. It did the thing the Grand Canyon does when you first catch sight of it from the window of a passenger jet. It's not like a mountain, or a mountain range. Even the Rockies only modulate the landscape—they don't interrupt it.

Now we saw that interruption, where the flat of the world fell away from the horizon. Where a crater had been punched through the face of the earth.

Terris swung us toward it. He circled, he rolled to one side, and we looked straight down onto the mine, onto its dozens of tiny yellow dump trucks. They drove along a curving network of dirt roads, through a mosaic of craters. Here they sped back to the hoppers, fully loaded and surprisingly fast, kicking up trails of dirt and dust. There, in the intimate cataclysm of a smaller pit, they waited in a group of two or three for their turn to approach a shovel, workers to their queen. And then away again, urgently, to deliver the next load.

The window pressed against my forehead. To the east and the south, I saw forest. But to the north, there was only the mine.

I wasn't horrified. But I had a funny feeling. Some kind of problem with scale. The trucks and the shovels looked so tiny—such toys and yet so huge. I had spent all week thinking about bigness, about weight, running through the synonyms for *huge*, and running through them again. The biggest machines in the world, they towered over a person with such magnitude and force. Now they were earnest beetles in a sandbox, themselves dwarfed by the vast footprint they were hollowing out.

"They look like ants!" Terris was shouting over the headset.

But they did not look like ants. They were too big to be ants. And some-how their very failure to be mere specks made them grow ever larger, and part of this growing was how much they seemed to shrink.

Vertigo rushed into the eye that tried to see it. And with the horizon circling around us, I knew that the mine itself, the panorama-swallowing mine, was barely a pinprick on the spinning body of the globe, and the globe itself a mote in the void, and the void itself a mote in another void, and I sat with my head pressed against the window—and felt, just a little, like puking.

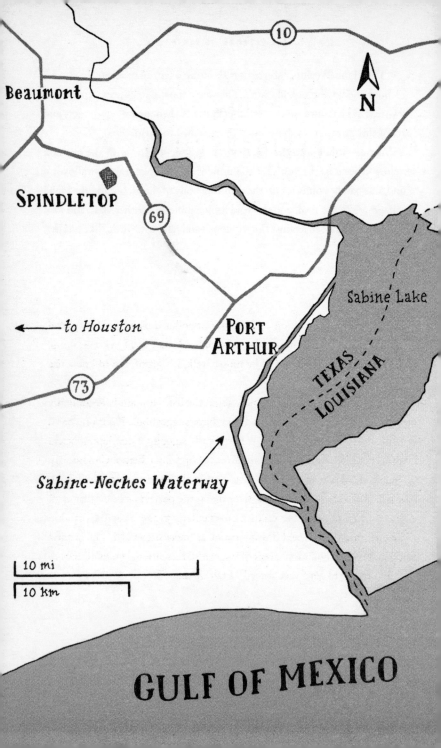

THREE

REFINERYVILLE

Tell folks that you're making a grand tour of polluted places, and they tend to get excited. A surprising number of people say they want to come along, and, although this turns out to be mostly talk, it's gratifying to know the market is there.

Most of all, people want to know about the list. How am I choosing my destinations? Based on what? And they have suggestions. Everyone has a favorite: a city that struck them as horrifically smoggy, a developing-world landfill they read about. Some make an easy leap from Chernobyl to Bhopal, taking up the theme of industrial disaster. But that doesn't seem quite right. And what if I want to check out a place that is the perfect embodiment of an environmental problem but that isn't particularly gross? Should I abandon it, just because I'm worried it won't count as "most polluted"? The criteria flood in: kinds of pollution, areas of the world, recreational possibilities . . .

"I'm trying to get a nice spread," I tell them.

\\\\\\\\\\

From Alberta, a powerful suction pulls south. And so they would like to build a pipeline. Another pipeline, that is—longer and better than what's already there. Leaving Canada, it would pass underneath the Alberta-Montana border and run clear through the heart of the United States to the Gulf Coast, ending at a clutch of refineries in Port Arthur, Texas.

Opponents of the Keystone XL pipeline, as it is called, argue that it would pose unacceptable environmental risks, even leaving aside the issue of how dirty oil sands oil is. The pipeline, three feet in diameter and buried underground, would transport diluted bitumen through such ecologically invaluable regions as the Ogallala Aquifer, which provides nearly a third of all groundwater used for irrigation in the United States, and is also a major drinking water supply. The threat to the Ogallala, the argument goes, is too great a risk to take. And then there's the question of whether the project would even be economically viable.

Pipeline supporters, on the other hand, claim that Keystone XL would be reliable and safe, and they contend that it would double the amount of oil sands oil that can be imported to the United States.

What Keystone XL definitely has going for it, though, is irresistible *symbolic* value. Judged by this admittedly dubious metric, a pipeline connecting northern Alberta and Port Arthur, Texas, is almost too good to pass up. Because if the oil sands represent the future of the oil industry, then Port Arthur represents its past, even its birth. And Keystone XL, should it be built, would physically link the two, feeding the future to the past, and tying the history of petroleum up in a tidy bow.

\\\\\\\\\\

They called it folly. To most people, it seemed ridiculous to imagine that there was oil waiting underneath the low hill known as Spindletop, near Beaumont, in Southeast Texas. But Patillo Higgins had been obsessed with it for nearly a decade. A local businessman and self-taught geologist, he had led multiple failed attempts to find oil under the hill, and still he persisted. The quintessential example of an entrepreneur driven beyond sound judgment, Higgins spent year after year chasing oil with nothing to show for it. He pursued his goal with a faith matched only by his own religious dogmatism, and even ceded ownership of his own company to attract new investors—all in an age when oil was used only for lamp fuel and lubricants. As a business plan, it was idiotic.

On the morning of January 10, 1901, Higgins wasn't even on Spindletop. Neither was his drilling contractor, a similarly obsessed, Croatian-born engineer called Anthony Lucas. They had no idea what was about to happen. Not even the drilling crew, as they ground the well deeper, past 1,100 feet, knew what they were about to unleash on Texas and the world. No idea that by lunchtime their well would be producing more oil than every other oil well in the country—combined.

It was the first *gusher*: the violent fountain of oil that in the old days would explode out of the ground when a new well broke through to a rich deposit. (Go see *There Will Be Blood* if you don't know what I'm talking about.) Nowadays, drillers understand how to control such things, but the gusher remains an archetypal American moment, as central to our folklore of wealth as gold rushes and tech IPOs.

Beginning on that January morning, the well called Lucas No. 1, or the Lucas Gusher, ran for nine days, spewing millions of gallons of oil onto the ground before it was brought under control. PURE OIL SPOUTING HIGH IN THE AIR—MUCH EXCITEMENT IN THE CITY ran the headline in Beaumont's *Daily Enterprise* on that first day. Just how much excitement can be traced in the work of the *Enterprise* headline writers over the following week:

January 12: MANY OIL PROSPECTORS ARRIVED TODAY.

January 14: FEVERISH AND EXCITED . . . BIG THINGS PLANNED WHICH WILL BE CARRIED OUT.

January 15: EXCITEMENT STILL HIGH. EVERYBODY GRABBING FOR LAND—PRICES SKY HIGH.

Their best effort, at once breathless and circumspect, ran on January 16: CROWDS STILL COME! . . . VARIOUS RUMORS OF IMMENSE TRANSACTIONS BUT VERIFICATION WAS NOT OBTAINABLE.

Within months, the population of Beaumont had quintupled; the sleepy town of Port Arthur, twenty miles down the road, was on its way to becoming a petrochemical mecca—and the Texas oil boom was on.

An oil industry already existed in the United States at the time. It had been built by John D. Rockefeller and his contemporaries, following discoveries made in Pennsylvania starting in the late 1850s. But oil had nothing like the dominance it has today. The internal combustion engine barely existed, plastic was decades away, and gasoline was considered an uninteresting refinery byproduct. Kerosene, the world's first bright, clean-burning lamp fuel, was the real game.

The Lucas Gusher produced more oil than anybody knew what to do with. Well after well was sunk into Spindletop in an orgy of drilling and speculation, and hundreds of new oil companies sprang up; you may recognize names like Texaco, Humble (now ExxonMobil), and Gulf (now Chevron). In Beaumont, the price of a barrel of oil dropped to below that of a barrel of water, so severe was the oversupply. Complicating this dilemma was the fact that this new Texas crude was ill-suited for making kerosene. Even if it *had* made for good kerosene, the writing was on the wall: kerosene lanterns were being replaced by electric lightbulbs.

The oil industry needed new markets. But what they eventually found— and founded—was a civilization. The dominoes began to fall almost immediately. First were the railroads: in 1901, the Santa Fe Railroad had a single oil-powered locomotive; four years after the Lucas Gusher, it was running 227 of them. Steamships in the Gulf of Mexico weren't far behind, changing over to fuel oil and lining up to take advantage of the glut. Mechanized agriculture and manufacturing took off in Texas, now suddenly the proving ground for the oil-based economy. Before long, the pattern was being repeated around the globe. Navies of the world switched to oil as well, signaling the abrupt geopolitical centrality of petroleum to the unfolding twentieth century.

And then there was the automobile, coming of age with eerie synchrony to the oil industry's burgeoning second wave. Several energy sources had been proposed for cars, among them electricity, but oil's new availability sealed the deal for the internal combustion engine. And the Texas crude refined nicely into gasoline. Before, gasoline had been considered a near-waste product; now it took its place next to fuel oil as the power source of the new age. It was time to pave America, and the rest of the world.

Over the following century, finding new markets for petroleum—new uses, new products, new *classes* of products—would prove to be one of the things that oil companies do best. And there is a direct line from the glut of oil on Spindletop to the omnipresence of petroleum today. As any oilman or environmentalist will tell you, oil seeps into every corner of our lives—our households, our economy, our politics. It fuels or abets almost everything we do, from tourism to warfare. I'm not telling you anything you don't already know. We live on oil, and by it, and its use is responsible for more than a third of global emissions of carbon dioxide, which, in an era of man-made climate change, is perhaps the most fundamental pollutant of all.

On Spindletop, though, on that January morning in 1901, all that was yet to come. Nobody knew that the twentieth and twenty-first centuries would be made of petroleum. And there had never been a gusher before. Nobody knew that a well could, without warning, explode into a glistening, green-black geyser. Nobody had ever danced in oil raining from the sky. When Lucas finally saw the roaring fountain that would immortalize his name, he just shouted, *"What is it?"*

\\\\\\\\\\

The late afternoon is a good time to drive to Port Arthur from Houston. You'll arrive at sundown, under a lavender sky deepening into purple, and see the distant lights and towers of a city, a wavering Manhattan spread out along the water, just where Texas decides it would rather be Louisiana.

What you see is not a city. Draw closer, and what you thought were buildings resolve into the spires and turrets of industry. They are refineries.

Soon you're surrounded. In one direction, there is water—in every other, the humming, roaring machinery of petrochemical digestion, a rusty Oz that churns through a million gallons of oil every forty minutes. It is from places like this that we receive our gasoline and jet fuel and plastic and everything else that we can't do without. Port Arthur is a refinery town, with oil in its veins, toluene in the breeze. It is the pungent center jewel in America's petrochemical tiara, also known as the Gulf Coast, a region that accounts for nearly half of the country's refining capacity. The US Department of Energy notes that the region has "the highest concentration of sophisticated [refining] facilities in the world."

Port Arthur, much like Fort McMurray, has a reputation as a shithole. But while the Albertans have managed to keep the oil sands mines at a discrete remove, Port Arthur is utterly dominated by its refineries, in ways that are impossible for even a casual observer to ignore. The downtown is literally encircled by steel forests billowing sulfurous air day and night. It smells like rotten eggs. Then there are the occasional *upsets*—accidents or malfunctions that sometimes result in the emergency release of fuel and other refinery goods into the atmosphere. The gases are burned off as they're released from tubes high above the plant, and people invariably describe refinery flares as awesome events, artificial auroras that paint the sky a glowing orange.

Most important, there are the habitual emissions of volatile organic compounds, things like toluene, benzene, and other contaminants that—it has been plausibly argued—result in elevated rates of respiratory disease, birth defects, and cancer for the communities that live with them. And once in a blue moon—seriously, only very occasionally—the plants self-annihilate. They explode. In Texas City, ninety miles to the west, a 2005 refinery explosion killed 15 people and injured more than 170.

The industry here is the direct legacy of the boom sparked by the Lucas Gusher, and the plants that overshadow downtown Port Arthur are the same plants that were built to receive Spindletop's oil, although a century's growth has transformed them. Valero (whose refinery first opened in 1901) and Motiva (1903) now cover almost as much land as downtown Port Arthur itself, and Motiva—in the middle of an expansion when I visited—is on its way to becoming the largest refinery on the continent.

Nevertheless, you can drive down Port Arthur's main street and fail to see another human being. With its rows of brick storefronts spread along a breezy coastal ship channel, downtown Port Arthur has the bones of a charming small city. But they are just that: the bones.

There are no grocery stores, no hardware stores—in fact I saw no surviving stores of any genre in downtown Port Arthur. There are no operating banks. Building after building sits vacant. Most are boarded up, burned out, or otherwise deserted. The industry that inhabits this city manages somehow not to sustain it.

As was traditional across America, the middle and upper classes of Port Arthur fled their city's downtown in the 1970s and '80s. Unlike in many other cities, though, the presence of the refineries has kept anyone with money from moving back. The result is a community that's among the poorest and most polluted in the nation—yet surrounded by multibillion-dollar companies. It's the perfect place to refine oil, incinerate toxic waste, and expand a petrochemical plant: a place where they're used to it. A place already so dominated by industry that nobody who matters will care.

The neighborhood to the north and west of downtown is poor and black. There are roofs still dressed with blue FEMA-issued tarps to cover damage from hurricanes of years past. I saw one FEMA tarp that had itself been repaired with another FEMA tarp. Beyond them towered the metal thickets of the refineries.

The best place to sit down for lunch in central Port Arthur—possibly the only place—is a soul food restaurant called Kelley's Kitchen. With its orange awning and hand-painted purple sign, it stands like an oasis among the vacant lots and boarded-up buildings. Inside, there is a single room with a painted concrete floor, a half-dozen tables, and a counter and stools in back. A young woman named Daisha served me shrimp, okra, and sausage over a pile of rice, with a pair of turkey wings and corn bread on the side.

Kelley's Kitchen was no mere restaurant. It was the latest venture from

Hilton Kelley, Port Arthur's leading environmental activist and all-around force of nature. Soul food is not typically a part of the environmental agenda, but Kelley took a holistic approach. "I'm about creating job opportunities," he said, as I buried my face in okra. "I'm about serving the community. I'm about encouraging young people to get business licenses, to do things that will help them get off the streets. 'Cause these streets will kill you faster than the pollution."

A tall, ample man in his early fifties, Kelley had an energy that was both generous and pugnacious. Above all, he was a man with hustle. When I first found him, he was sitting at a table working on his laptop while eating lunch; moments later he was outside with a crew of helpers, hauling a pair of heavy wooden stalls to a spot in front of his restaurant. In preparation for the upcoming Mardi Gras parades—the only time of year when central Port Arthur sees some life—Kelley was planning to sell "food and hats and what-not" to passersby. Moments after that, we were back inside the restaurant and Kelley was pointing out the new dance floor, off to one side. "I love dancing. That's why I built me a dance floor." He was an experienced carpenter, and power tools littered the cab of his pickup truck.

But above all, he had devoted himself to picking environmental fights in Port Arthur. His organization, the Community In-power and Development Association, had recently blocked the importation of PCBs from Mexico to a nearby incinerator. It had also fought the Motiva refinery expansion, holding it up and forcing concessions from the company on monitoring and community investment.

Kelley was also working with a group called the Southeast Texas Bucket Brigade, doing grassroots air-quality measurement, in hope of filling the massive gaps in monitoring left by industry and government. The figures available for refinery emissions, one environmental lawyer told me, are based not so much on actual monitoring as on calculations made by the EPA—calculations that can be decades old. As a result, it's nearly impossible to know exactly what's drifting out of a refinery in any given week.

"Toxic exposure!" Kelley said. "You've got hydrogen sulfide. Benzene, a known carcinogen. Thirteen butadiene. Occasionally, you've got explosions that will rattle your windows. Some people are living with storage tanks sixty feet from their backyards. If one of those things went up, it would incinerate everything within a quarter block." He had strong words for the state regulators—"They have to actually do their jobs!"—as well as for the Environmental Protection Agency, and before I knew it, he had become a one-man poetry slam, performing a piece called "My Toxic Reality," written after he'd spent a sleepless night listening to his house being rattled by a nearby refinery flare.

In Kelley's pickup truck, we rode slowly through West Port Arthur, taking what he called his "toxic tour" of the city. Until 1965 or so, he said, segregation meant that African Americans weren't allowed to live anywhere but the West Side. It was no coincidence that this was the part of town closest to the refineries, hemmed in by Valero and Motiva.

As we drove, Kelley told me his life story with the fluency of someone used to talking to journalists. He grew up in Port Arthur in the 1960s and '70s, then joined the Navy and ended up in California, where he became an actor and stuntman. In 2000, he came back to Port Arthur for Mardi Gras and was shocked by the poverty and hopelessness he found.

"I would take these little walks," he said. "And I started wondering, what the hell happened?"

He decided to move back, hoping to find some way to help, and soon found himself focused on the local environment: lobbying for better monitoring and enforcement, and standing by the refinery gates with signs demanding change.

"I thought I'd be here two or three years when I came back," he told me. "Now it's been ten years, and I don't see no end to this environmental fight."

We drove on, heading along West Seventh Street, the artery running from downtown, through the poor neighborhoods, toward the bridge that crosses the ship channel. "People are just appalled to even drive through here," Kelley said. "They talk about building another bridge, just so people don't have to drive down Seventh Street, so they don't have to go through the West Side."

He told me it was part of a larger pattern—a conspiracy, even—that threatened to starve West Port Arthur out of existence. "I think a plan was developed," he said. "A sinister plan. I don't have any proof, but I'd stand up and say that in front of anybody. You have a community with a thirty-billion-a-year company on one side and a forty-billion-a-year company on the other side, and yet it's one of the most dilapidated communities in Texas. It don't add up."

Driving past the football field of the deserted former high school, Kelly pulled the truck over. He was looking in the rearview mirror. He had done this more than once during the tour, letting people pass us as we crept around the neighborhood.

"Come on, drive around me!" he said. Finally the car passed us. He watched it go.

"I'm real leery about people following me," he said. "I wouldn't say *paranoid*. I'm cautious. And of course, I've always got my little friend."

He pointed at a small soft case resting between us on the floor of the cab.

"Oh," I said. "You mean—"

"That's right," and then he was holding it up, a heavy piece of metal that looked very much like a handgun.

The afternoon had taken a turn. "I keep it loaded," he said. "And one in the chamber." He said he carried the gun partly because of the crime rate in Port Arthur—but only partly.

"There are some people here who hate my guts," he said. "They think I'm a troublemaker. That I'm going to make them lose their jobs. But I am not trying to shut the refineries down. I just think they need to abide by the regulations we already have. By the Clean Air Act. And they're not."

A phrase like "abide by the Clean Air Act," I noticed, took on a nice urgency when you waved a loaded .40 caliber around while saying it.

We passed by a storage yard full of components for the Motiva expansion. Kelley was talking about the products that came from the refineries. He knew they were important. He knew we all used them. He was, after all, driving a truck that probably got about fifteen miles to the gallon.

"My campaign has always been, it could be cleaner," he said. "It could be done safer. Our health could be protected. The companies should open up. Let us know what's going on. Let us make informed decisions." We made a pair of right turns onto roads flanked by pipelines.

"In fairness," he said, "they're doing a little better."

He stopped the truck. We had come to the Carver Terrace housing projects, a set of two-story brick buildings facing the Motiva refinery. Kelley pointed down a pathway.

"I was born right in there," he told me. "First floor."

He twisted around in his seat and pointed at a small, deserted playground across the street. "That old swing is the one I used," he said.

Several hundred yards beyond the playground were the storage tanks of the Motiva refinery, and beyond them the refinery itself, a jungle of pipes and towers, steam plumes and winking flares. The breeze carried a rancid aftertaste.

"We would breathe this air," said Kelley, staring at the refinery. "We used to joke about it. My mother would say, 'That's money you smell.' And we'd say, 'No, that's death!'"

"I guess it's both," I said.

He sighed. "Yeah. It's both. But it wasn't *our* money."

It was an irony of Kelley's work. With one breath he called the refineries a "cancer" that needed to be cut out of the city, and with the next he lobbied for their owners to hire more locals.

"Look which way all the traffic's going," he said as we passed the gate of the Motiva plant. It was the end of the shift, and all the cars were headed out of town. "These people work here, but they don't live here." Kelley wanted jobs for West Port Arthur. If it was going to suffer the refineries' effects, shouldn't it also share in the wealth? In an area so dominated by industry, half the point of environmental activism was just to get a piece of the action.

\\\\\\\\\\\

On a bright weekend afternoon, I went for a run. Valero shone in the sun as I approached it at a blistering saunter. Seen like this, with time to look, it was somehow hypnotic in its tangles of silver and rust, its smokestacks and flares and steam plumes. Deep inside its chambers and towers, the entire roster of hydrocarbons was dividing itself into fractions of kerosene, gasoline, and jet fuel, and being cracked and catalyzed, cousin by cousin. I peered sweatily at the atmospheric distillation unit and the vacuum distillation unit as I passed, at the catalytic cracker and the hydrocracker, at the hydrotreater and the coker, at the catalytic reformer—not that I knew which was which.

No less than with Chernobyl, it is excruciatingly difficult to make definitive statements about the health effects of Port Arthur's environment. But there is at least one clear effect, which is that many people here—not just the environmental activists—simply assume the worst.

A taco truck was parked at the southeast corner of the Valero plant, just outside the fence from the resplendent steel sphere of a storage tank. The truck's owner was a genial Mexican immigrant who told me he had seen the plant release flares so large that he could feel the heat on his back, even here outside the fence. Through the window in the truck, I asked him if he thought the air from the plant was bad.

Of course it's bad, he said. It smells terrible. *Feo* was the word he used—Spanish for "ugly." You get all kinds of things from that air, he said. Cancer.

When I suggested that he find some other place than the Valero fence line to park his taco truck, he laughed.

You've got to make a living, he said, and handed me a taco, *al pastor*, on the house.

Then there was Ray, a refinery worker who struck up a conversation with me at a bar downtown. He had worked at the BASF petrochemical plant for twenty-two years.

"Lemme tell you something," he said, drunkenly waving a plastic cup of Boone's Farm. "By the time I'm fifty, I know—I don't guess, I *know*—I'm gonna have some kind of cancer. Everybody at that plant knows, beyond a shadow of a doubt." Ray was also of the opinion that a terrorist attack on one refinery could lead to a catastrophic chain reaction that would level fifteen plants between BASF and downtown. "This place is a time bomb," he said with some joviality.

In Port Arthur even the most ardent civic booster may shift seamlessly onto such topics. Five miles north of downtown, at the convention center, I met Peggy and Laura, two friendly ladies in charge of the Majestic Krewe of Aurora's annual Mardi Gras Ball. Peggy was such a loyal daughter of Port Arthur that she was still nursing a grudge against Janis Joplin (who grew up here) for once having talked trash about the local high school. But I barely had to let it drop that I was a writer interested in the environment before Peggy took up the cudgel.

"Cancer!" she exclaimed. "We've got lots of cancer around here. It's the refineries. And the incinerator. You know about the incinerator, out by the highway? Where they're burning all that nerve gas? Why, they burn all kinds of horrible things out there. That stuff is going to get into the aquifer," she said. She sounded almost proud.

But I wasn't here to follow cancer down the rabbit hole. I could have spent a lifetime trying to nail down what portion of the city's elevated cancer load was real and what was merely assumed—not to mention the health effects of a citywide *assumption* of cancer. Leave it to the epidemiologists. What I wanted to see was how the landscape and culture of Southeast Texas had been shaped by more than a century spent as Big Oil's ground zero. An economic and cultural ecosystem of sorts had been created when the Lucas Gusher spat itself onto the earth, one that persisted to this day.

"Would you like to come to the ball?" Laura asked. She had tickets in her hand.

\\\\\\\\\\\

By the time I returned to the convention center the following night, it had been transformed into a fantasy of glitter and noise. Smiling men and women wearing tuxedos and evening gowns flowed by in a cackling stream, bringing a palpable enthusiasm to the project of getting drunk.

My friends Scott and Lorena had come out from Houston for the occasion, and although we had tried to spruce ourselves up, we stuck out. It turns out there is no way *not* to stick out in a convention center full of people dressed as harlequins and playing cards. This year's theme was "The Games People Play."

It was a party fueled by beer and oil. The projected logos of its sponsors bejeweled the ballroom walls. Both Budweiser and Bud Light were represented, as well as the Valero Port Arthur Refinery, Total Petrochemicals, BASF, Sabina Petrochemicals—all the major players. They were here to celebrate with the city's upper crust, the inheritors of the economy created on Spindletop. People who I doubted lived in West Port Arthur. Dance music pounded from speakers hanging overhead. Green lasers shot out over the crowd from the stage, tracing twitching planes in the fog-machine atmosphere. It was hard not to think of the "feverish and excited" scene described by Beaumont's *Daily Enterprise* in the first weeks of 1901. I turned around to see a young woman in an elaborate Cinderella costume. The Queen of Diamonds? Then Scott was there, holding three aluminum bottles of Bud Select.

"You must not miss the tableau," Laura had told us. And now it had begun, an elaborate ceremony that was most likely descended from pre-Columbian human sacrifice rituals, and that had now been retasked for the apportionment of social standing among high-status members of the Krewe. To validate this status, chosen individuals would appear in male-female pairs, draped in gaudy costumes conforming to the ball's theme—in this case, games. Duly announced, the couple would then parade around the ballroom on small chariots pulled by young men in maroon vests.

The first couple appeared. I don't recall whether they were dressed as Yahtzee or as craps, only that the man was equipped with a large, feathery headdress and a suit of blazing sequins, and the woman with a massive corona of flowered ruffles. The couples kept coming, each dressed as a board game or a card game or a game show. It took hours. The crowd thronged around them, a riot in formal wear, waving madly to catch the plastic beads and party favors being thrown by the couple of the moment, who would eventually ascend to side stages where they would pose for the remaining duration of the tableau, feathery demigods on display.

Motiva was in the house. Soon to be the largest refinery in North America, it had sponsored a couple dressed as the board game Mousetrap. After seeing the snaking insanity of the refinery itself, it seemed almost too good to be true that Motiva would come to a party dressed as a Rube Goldberg machine. I got up from our table to get a closer look. Lady Gaga beat her fist against my chest. A quartet of dancers gyrated across the stage in the distance. Small Frisbees with blinking LED lights flew in parabolas over the crowd. The Motiva queen showed her teeth to the ceiling. Beads exploded from her hands, filling the air with plastic shrapnel. Through the haze, I saw the silhouette of a young man in a perfect cowboy hat, his profile seething in the flare of a spotlight.

Scott and I found Laura on one of the side stages, utterly transformed from the day before. Then, she had been a short, unprepossessing woman in jeans and sensible shoes. Now she was dressed as *Wheel of Fortune*, a Pat Sajak fever dream of sequins and feathers, with an enormous model of the wheel rising from her shoulders. She was ten feet tall, an Aztec high priestess of TV game shows, with a floppy BANKRUPT wedge running down her leg. One of the first out of the gate, she had been standing in presentation for upward of an hour, next to a nebula of plumage that was a woman dressed as Monopoly.

Beneath her towering outfit, Laura's smile had frozen into a rictus of determination. I was concerned she might collapse.

"You look amazing!" I shouted over the music.

"Thank you!" she screamed.

"I don't know how you can stay on your feet with that costume!" I said.

"It's much lighter than it looks!" she warbled, and took a swig from a bottle of water.

The tableau was reaching its climax. Shafts of light exploded from a giant mirror ball. Laser-light unicorns galloped across the back wall of the ballroom. A king and queen were announced, and all hell broke loose. Confetti swirled in drifts. A conga line fought its way through the hurricane. An elderly woman danced alone in circles, her arms raised in triumph, or surrender.

\\\\\\\\\\

Within two years of the Lucas Gusher, overdrilling bled Spindletop dry. The rush was over—or rather it moved on, spreading out to new oil strikes elsewhere in the state and country. Later, in the 1920s, a new wave of exploration led to a second boom on Spindletop. Then, in the 1950s and '60s, the land was mined for sulfur and salt brine, causing the ground to subside in broad depressions, as if letting out a great sigh of geological exhaustion. The forest of derricks was long gone. The place was left empty. Today it is a range of sand and scrub, dotted with the wreckage of oil production past.

On the south side of Beaumont, between Highway 287 and the Lamar University driving range, I went looking for Lucas No. 1. It was raining when I got there. On a wide, soggy lawn, a stone obelisk stood cold and lonely in the damp. I read the engraving on its base:

ON THIS SPOT, ON THE TENTH DAY OF THE TWENTIETH CEN-

TURY, A NEW ERA IN CIVILIZATION BEGAN.

But someone should carve that obelisk a footnote. This was not, in fact, the spot where it all happened. The obelisk had been moved from the original site when the ground began to subside. This was merely the front lawn of the Spindletop–Gladys City Boomtown Museum.

It's not a bad museum, otherwise. They have built an entire replica boomtown village, and next to the obelisk, there is a life-size replica of the Lucas No. 1 oil derrick, fitted with a large nozzle, as if from a fire hose. For a hundred dollars, I was told, you can have this nozzle turned on, and it will spray water at the same pressure and to the same height as the original Lucas Gusher. Oil companies sometimes bring new hires there to celebrate.

As for the actual Lucas Gusher, it's about a mile south of here, on private land. The Spindletop oil field has been designated a national historic landmark, but it's also designated *Authorized Personnel Only*.

\\\\\\\\\

The oil that once came from Spindletop now comes from more remote oilfields, or from offshore wells in the Gulf of Mexico, or is imported by tanker from overseas. One day it may come, by pipeline, from Alberta. In any case, the refineries of Port Arthur are tied less to the people living outside their fence lines than they are to the distant sources that keep them humming.

But in Southeast Texas, oil sustains more than refineries. Its nourishment spreads out through circle upon circle of lesser players that cluster and compete at the oasis of its wealth, living off its power and success—and even off its disasters.

On January 23, 2010, an oil tanker called the *Eagle Otome* entered Port Arthur's ship channel, the Sabine-Neches Waterway, with 570,000 barrels of crude oil on board, destined for the ExxonMobil refinery in Beaumont. To make its delivery, the tanker would have to transit the length of the ship channel, a thin, man-made strait that runs inland from the Gulf, along the frontage of downtown Port Arthur, and then up toward Beaumont. The channel measures not even three hundred yards wide at points, and the navigable waterway—the part deep enough for ships—is even narrower. It is a hard needle for any large vessel to thread, and the *Eagle Otome* was more than eight hundred feet long.

While the age of gushers is long past, there is still occasion in Port Arthur for the unplanned flow of petroleum. As the *Eagle Otome* came around a mild bend in the channel, it swerved off course, fishtailing slowly down the channel as it approached the wharf in downtown Port Arthur. The tanker—nearly as long as the channel was wide—skewed across the waterway, colliding with a vessel tied up at the wharf and obstructing the path of an oncoming towboat. The towboat, pushing a pair of 250-foot-long barges, had no choice but to plow directly into the *Eagle Otome*, ripping open a neat gash in the oil tanker's hull. In what seems like a great stroke of luck, though, only 2 percent of the tanker's oil spilled through the opening.

On the other hand, we're talking about 2 percent of more than 23 million gallons of cargo. It was the largest oil spill Texas had seen in two decades.

\\\\\\\\\\

As with an oil find, so with an oil spill: for as long as it lasts, it is a source of work. On Spindletop, that meant on and off for decades. In the case of the *Eagle Otome*, it meant a little over two weeks. There were cleanup companies to deal with the spilled oil, and tugs to tow the damaged ships away for repairs. There was the media, trying to puzzle out the causes of the accident and covering the closure of the channel. A more catastrophic incident might have sustained them for a month or more. (As for me, it was just dumb luck that I happened to show up in Port Arthur only a couple of weeks after it happened.)

An oil spill is a boon of sorts even to environmental activists, whether as additional motivation or as convincing, public proof of an issue's importance. The threat of poisonous hydrogen sulfide gas from the spill prompted a short evacuation of downtown Port Arthur—a fact that had already become another arrow for Hilton Kelley to shoot at the refinery companies.

It might not be the most efficient way to extract value from oil, but the fact remains that a spill is not only a spill. It's a massive carcass, which we gather around to eat.

At my hotel, the parking lot was crowded with trucks bearing the logos of companies like Clean Harbors and Oil Mop LLC. I was not the only one who had chosen the Ramada: the Coast Guard had set up its spill response headquarters in one of the conference rooms. Khaki-wearing men strode in and out of the lobby with an air of can-do seriousness. At its height, the cleanup had put something like two thousand people to work, but now things were winding down, and the mood was almost festive.

"I hear you're leaving us," said the hotel manager to a passing cleanup contractor.

"Well, maybe we'll be back," said the contractor.

"For the next oil spill?" called a woman from behind the check-in desk.

In the empty hotel restaurant, I met with Jeremy Hansen and Bryan Markland, two well-scrubbed Coast Guard officials working on the cleanup effort. "You've got all these local cleanup contractors poised to jump," Hansen said. "It's cutthroat."

Markland told me that cleanup contractors often begin their work even without being hired, confident that if they do the work, someone will have to pay for it. And so skimmer boats materialize, hungry for oil, and lines of floating containment boom sprout to cordon it off, and the cleanup's economics bloom.

It is discouraging, though, to reflect on how little even an effective cleanup can achieve. "Most oil spills, if you get more than 15 percent of the oil recovered, you're doing good," Markland said. "We think we're up in the 30 percent range on this." The rest of a spill, he told me, simply evaporates or disperses to what he called an *unnoticeable sheen*. Which is to say, most of the cleanup is actually done by nature—or isn't done at all.

Hansen was sitting back, his arms crossed. He looked a little mischievous.

"Did you see the Port Arthur slogan?" he asked.

I laughed. I *had* seen it, on the website of the Port Arthur Chamber of Commerce. It might have been the most ill-advised civic motto of all time:

Port Arthur: Where Oil and Water Do Mix. Beautifully.

Hansen smiled and shook his head in disbelief. "It's a good thing they don't," he said. "Or it would be a lot harder to clean up."

Then there was Rhonda, the grumpy pelican lady. She was in charge of rescuing and rehabilitating birds oiled by the spill. A bustling woman in a salmon-colored shirt exploding with pockets, she struck me as deeply unsentimental about her work, and she didn't hide her annoyance that I was interested in it. Had I been naive to imagine that the bird savior of record would share a little enthusiasm for bird saving? But Rhonda was no simple bird lover. She was the director of Wildlife Response Services LLC—just one more contractor providing post-spill services.

"What is it you want, exactly?" she asked.

Eventually she resigned herself to my presence, and soon we were standing in the corner of a cavernous warehouse, staring at a pelican. Miraculously, only nine birds had been oiled in the spill: a loon, a cormorant, a seagull, a spotted sandpiper, a black-crowned night heron, and four pelicans. With one exception, they had all been released back to the wild after being cleaned, fed, and housed until they were back in fighting form. The lone holdover was a brown pelican now living in a plywood pen with a sheet over it, in a temporary rehabilitation center downtown.

A rehab worker raised the corner of the sheet and the three of us peered through the narrow opening. I held my breath. Inside, lit with the radiant orange glow of a heat lamp, the single pelican sat motionless on a low perch, a Buddha with folded wings.

"He'll puke if you pick him up," Rhonda said. She was advising the rehab worker not to let the pelican take a test swim yet. "You can't mess with them when they eat."

It had been a rough century for pelicans on the Gulf Coast. A hundred years earlier, fishermen had gotten the idea that pelicans were competing with them, and had slaughtered them wholesale. Worse still, by the 1950s, our release of pesticides into the environment had become a two-pronged machine of pelicanic destruction: DDT weakened their eggs, killing chicks before they even hatched; and Endrine killed off the fish that were their food, starving pelicans en masse. By the late 1960s, they had almost completely disappeared from the Texas and Louisiana coasts.

The late 1960s and '70s saw pelicans reintroduced from Florida, and a ban on the persistent organic pollutants that undermined their niche in the ecosystem. Today, the coast is once again crawling with them. Which is not to say they are invulnerable, even without oil spills.

"We have a pelican die-off every year," Rhonda said as the rehab worker closed the sheet over the pen. "There are some pretty harsh cold snaps. The fish move off, and the birds don't get enough food." She shrugged. "I don't know, I'm not a biologist."

Then she laughed. "These guys were actually lucky they got oiled," she said. "They've been fed quite well."

\\\\\\\\\\\

The Hotel Sabine is the tallest building in Port Arthur, and the best vantage from which to watch the aftermath of an oil spill. There's simply nothing more pleasant than to book a south-facing room on an upper floor and enjoy a gimlet as the cleanup workers buzz up and down the waterway.

At least, it *would* be pleasant. The Hotel Sabine has been abandoned for years, and now stands vacant and eyeless, not only Port Arthur's most prominent landmark but also its most obtrusive eyesore. Unless you plan on breaking in, there will be no tenth-floor views of the ship channel for you.

Instead, I drove down to Pleasure Island, the grassy artificial landmass on the other side of the channel, to watch the men in Tyvek wrap things up. Oil Mop boats dragged lines of floating containment boom up and down the waterway, their hulls smeared brown with crude. The *Eagle Otome* had already been spirited away for repairs, and the mood was calm—pastoral-industrial.

The channel's surface was unremarkable from a distance, but closer inspection revealed that a not-yet-unnoticeable sheen of oil persisted near the shore. I crouched on a sloping concrete slab that formed part of the bank and watched the filmy rainbow burble over the rocks.

There was a man standing on the bank just up the channel. He was short, with blue-tinted glasses and a suede cowboy hat jammed down on his head. And he was fishing.

His name was Nelson. Originally from El Salvador, he said he had been in the United States for ages. He owned a dump truck in Beaumont and made his living hauling dirt and gravel for road construction jobs. In a drawl that was half Texas and all Salvador, he told me this was his favorite spot to fish.

"Last weekend, they had that spill?" he said. "I show up here, a lot of oil. A *lot* of oil. I went further up the channel. Where it was clean."

We looked at the edge of the channel below our feet, where the waterline curled in colored wavelets of petroleum.

He frowned with approval. "Today, though . . . I think is okay."

"It doesn't bother you at all that there's still oil on the water?" I asked. "I mean, there's still guys in orange suits."

"No, man!" he said, and waved at the channel. "If you fish like this, with some oil there, then you don't have to use no oil when you cook it!" He cackled. "That's a joke."

He had extra fishing rods. I probably hadn't fished in twenty years, but it came back after a pair of somewhat hazardous casts, and soon we got on with the business of letting the crabs of Port Arthur steal Nelson's bait from our hooks. The Oil Mop boats continued their rounds, and Nelson cracked open his supply of Coors Light.

He seemed glad to have me there, and soon we were talking about his divorce, about how much he missed his sons. He told me he wanted to find a girlfriend from overseas, and about his complicated attempts to find one over the Internet. It sounded less like online dating and more like a Nigerian banking scam, but that didn't seem to bother Nelson.

What about you, man? You got a girlfriend?

I told him I did.

As a matter of fact, I was engaged. The Doctor and I were getting married. And once we were married, we were going to India, to take the world's first pollution tourism honeymoon. That she considered this even tolerable seemed like further proof of true love. Cruising the world's most degraded rivers, just the two of us . . . I was pretty sure it was going to be more romantic than it sounded.

There was a tug on the line. I did as Nelson had taught: I pulled up sharply on the rod to set the hook—and waited. "You feel something again after that, you've got a fish," he'd said. But so far the tugs on my line had signified only that my hooks were now empty of bait.

But this time there was another tug on the line, and another—an irregular rhythm drumming against the rod and reel. I started reeling, and like magic, two large fish appeared in the water.

Two.

Nelson threw down his rod, whooping. "Pull him in!" he cried. "Pull him in! You got two!"

I pulled and reeled and yanked the fish toward the bank, where Nelson grabbed the line and pulled them out of the rainbow-stained water, beaming at my success. The fish hung from the line, exhausted and gaping, each of them a good sixteen inches long. They were the largest fish I had ever caught. Larger, perhaps, than any fish ever caught in the history of the world.

"That's called drum," Nelson said. *Drum.* I had caught oily drum. He slapped his leg. "That's going to cook up real good!"

\\\\\\\\\\

Rhonda was on the phone. They were about to release the pelican. Over the line, I could hear her teeth grinding. She hadn't wanted to make the call, but I had put in a request with the Coast Guard to ask her to.

She sounded hopeful that I wouldn't be able to make it, and gave me only very vague directions. Her team was already on the road, she said. It was probably too late for me to find them.

But if she thought she could hide this pelican release from the world, she was mistaken. I sped across town, crossed an imposing cable-stayed bridge over the northeast elbow of the ship channel, and then doubled back to the south. Pavement turned to gravel, and the road plunged into a wetland park, stands of grass interlaced by channels of placid water. To the west, the horizon was decorated with the distant skyline of the refineries, tiny thickets of smokestacks and fractionating columns.

Driving south, I passed the occasional clot of trash—a shattered television on the shoulder, a pink recliner submerged to its forehead in a placid side channel. Cormorants and pelicans wheeled by, and cranes and herons, and other long-necked beasties. Here and there, men sat by their pickup trucks and fished. The fish were not biting, they told me.

Of course they're not biting, I thought. *You're fishing in clean water.*

Finally I spotted a pair of SUVs parked by the canal that ran parallel to the road. It was Rhonda's crew. I had caught them in the act.

The pelican was already in the water, floating next to the reeds on the far side of the channel, maybe fifty feet away. I walked up to Rhonda and her three colleagues. She registered my presence with obvious disappointment. The rehab worker from the warehouse was there, too. "Hey, buddy!" she said, proving that not all pelican ladies are grumpy.

We watched the pelican. There was an air of expectation, even concern. "C'mon!" someone said. "Fly!"

But the pelican did not fly. It merely floated. And the longer it floated, the more tense everyone became. At last, it dunked its head and unfurled its wings, and, with a broad flap, splashed itself with water. The crowd broke into applause.

"Yes!" said Rhonda. "That's what we're looking for!" She took some pictures. "Do that again!" she shouted at the pelican, and it obeyed, stretching and flicking its wings over and over, bathing in the churning spray, improbably majestic.

Rhonda turned to me. "See?" she said accusingly. "It's not very exciting."

"It *is* exciting!" I protested. I couldn't span the absurdity of not being able to convince a rescuer of wildlife that wildlife rescue was, in fact, interesting.

Rhonda turned back to the pelican, now swimming in idle circles, and began screaming at it.

"STAY AWAY FROM PEOPLE!" she bellowed. "FLY OFF INTO THE BUSHES! STAY! AWAY! FROM PEOPLE!"

She caught her breath. "That's the problem, is if he got used to people."

"He's gonna miss that heat lamp tonight," someone said. The forecast was calling for cold weather. "He's gonna wish he were back in that warm cage."

"No," said Rhonda. "He hated it in there."

\\\\\\\\\\\\

It heaved toward us: a mountainside of black steel. I was standing on a gangway, clutching the rail as our boat rocked and turned. I was facing port. That means left. It was hard to look anywhere else; the thing approaching to port had no end. It spread up and out from the water, an endless wall of rust-streaked metal, and we were falling toward it.

Duane was there, the trim Boy Scout of the sea, wearing a backpack.

"Don't let go of the railing until you have a good grip on the ladder," he said. I made a noise like a strangled fish.

The tanker was so tall and so wide that it seemed to outstrip my entire field of vision. Yet the distance between it and us was surprisingly nimble in the way it diminished. *At this rate*, I thought—

Then we were at the ladder, a wooden ladder hanging down the rain-soaked hull. *Wood?* Its treads hung from thick ropes dark with sea scum. I grabbed it and found myself clinging to the outside of twenty million gallons of Mexican crude. We had boarded the *Pink Sands*.

When Port Arthur began its life as an oil town, ships came here to take the stuff away. But now, of course, they bring it in, by the half-million-barrel load. The question, especially pointed in the aftermath of a spill, is how to make sure these ships don't crash, despite taking so much cargo up such a narrow waterway. Or perhaps the question is why they don't crash more often. The answer, I was here to learn, is that any large tanker that enters the Sabine-Neches Waterway is required to carry a pair of Sabine pilots.

On the wide, linoleum-floored bridge, we met Captain Tweedel, Duane's colleague and president of the Sabine Pilots. A tall, clean-cut man wearing chinos and a braided belt, Tweedel had grown up in Port Arthur. (Though he now lived in Beaumont. His wife had insisted on not living in sight of a refinery.)

The two captains got down to work, staring out the window with that look people get when they have just taken control of fifty-five thousand gross tons.

"Full ahead," Tweedel ordered.

"Full ahead," said the helmsman.

I was on my second visit to Port Arthur, several months after the *Eagle Otome* oil spill, and the channel had long since returned to normal operation. But questions still lingered; the government had yet to finish its investigation into the cause of the accident.

In the meantime, the Sabine Pilots had begun working with a public relations consultant, and were surprisingly willing to let me tag along. They wanted their story to get out.

The trick to keeping an oil tanker from crashing and spilling oil all over your ecosystem, it seems, is to have Charlie Tweedel and Duane Bennett standing on the bridge. They stand there, staring out the window, at a piece of water they have studied and navigated for years, and occasionally tell the nice Filipino man at the helm to adjust the rudder by ten degrees. It's more complicated than that—but not by much.

"Port ten!" said Duane from the captain's chair. He looked like a nicer, nerdier Captain Kirk.

"Port ten!" came the response. Nothing happened. The deck continued to vibrate with the power of the engine. Then, six hundred feet in front of us, the nose of the tanker began creeping to the left.

Hardly any major harbor or channel lets ships enter without a local pilot aboard. The stakes are simply too high—and the navigation too tricky—to leave it to some guy who doesn't know the route's every curve and shoal. The pilots meet their charges in open water, before the ships enter the channel, and clamber aboard—pirates by invitation. Tanker captains are more than happy to hand over the controls, as they must.

"We consider ourselves as the buffer, as protection to the environment," said Tweedel, staying on message. "The government expects us to act to protect the waterway and the populace from some radical conflagration or pollution."

"And the accident in January?" I asked.

"I don't want to talk about that much," he said. "It's still under investigation." He told me there was no single factor the accident could be hung on.

We slid forward through the cold, misty morning, passing from the outer harbor into the green mouth of the channel. Idle oil platforms lingered against the bank to our left, waiting for contracts or to be torn apart for scrap. On the navigation table, I had seen a map of the coast, marked with dozen upon dozen of offshore oil wells, punctuating the Gulf with surprising density. "They're like fleas," Tweedel had remarked.

Port Arthur's ship channel is not only so narrow that two large tankers going in opposite directions would have no room to pass each other, but also so shallow that Tweedel described it as a "muddy ditch." He told me that, at the moment, we were drawing thirty-nine feet. That meant the bottom of the hull was riding thirty-nine feet below the surface of the channel.

"What's the maximum draft you can have in the channel?" I asked.

He smiled. "Forty."

"Midship!" shouted Duane.

"Midship!"

The task of piloting a tanker requires continuous attention. "As a pilot, you'd really be taking a risk to leave the helm for more than a minute or two," Tweedel said. He pointed at an oncoming barge. "If he ran aground, I'd have to immediately take action. And I've seen those guys run aground lots of times."

"We're compensated for the risk," said Duane. Piloting paid well.

Tweedel peered out at the low, misty sky. It was also up to the pilots, he told me, to stop tanker traffic in the channel if visibility was too poor. Today's conditions were just good enough.

"It gets foggy for three or four days, and people start screaming for their crude oil," he said. If the supply of oil didn't keep up, the refineries might have to lower their production—and that would cost them money. There was huge pressure on the pilots to keep traffic moving.

"We want to support the industry guys," Tweedel said, "but we don't answer to Motiva or Total."

We slid forward, an impossibly great momentum, a floating machine literally as long as a skyscraper is tall. I looked over at the helmsman. He was holding a semicircular wheel not unlike the steering wheel of a go-cart. It seemed like it would be very easy, had I wanted, to shove him aside and twirl that wheel, and create a new round of honest work for nearly everyone I had met in Port Arthur.

"Starboard twenty," said Duane.

"Starboard twenty," said the helmsman.

Starboard? Earth to Duane! *Starboard?* I would have said we needed some port rudder, if anything.

"Midship," said Duane.

"Midship," said the helmsman.

And with that, subtly, our leviathan shifted its attitude and slid true, perfectly congruent to the grassy shores of the channel.

Duane handed the command off to Tweedel and walked over to the window. I told him I had been playing a game called Drive a Supertanker, and losing.

"It's more art than science," he said. "You have to know the science, but there's a feel you get. If you can't feel the vessel, you won't be good as a pilot."

He took my notebook and started drawing diagrams, explaining the hydrodynamics of a large ship moving through a narrow channel. The size of a ship affects how it handles in such a limited space. As the ship comes closer to the side of the channel, the water being displaced by the vessel creates pressures and suctions that interact with the narrowing space between the ship and the bank. The ship begins to handle differently, steering itself, resisting in ways it wouldn't in open water. These effects not only constrain how the vessel can be piloted, and how quickly, but also allow the person in control to sense the ship's position in relation to the channel, based on how it's handling.

"A ship is a totally different animal in these channels," Tweedel offered.

Barges passed us coming the other way, carrying refinery products, wood chips, grain. We were the only large tanker; because the channel was too narrow for two such ships to pass each other, their comings and goings were scheduled so that it never happened. But even the movements of smaller craft had to be carefully coordinated to ensure safe passage through such constricted waters. So Tweedel and Duane were also traffic controllers, scrutinizing the approach of other vessels, ordering them around, negotiating what maneuvers they and the *Pink Sands* would take as they met.

"I'm gonna need some of that water, Cap'n," Tweedel said over the radio, cajoling an oncoming tug into position.

We were entering Port Arthur, passing under the soaring eyesore of the bridge that connected Pleasure Island with West Port Arthur. The Valero refinery crawled by on the left, superb in the mist. The Sabine Pilots should charge for tours of the waterfront. Throw in a bottle of champagne and some strawberries, and nobody would ever have to ride in a hot-air balloon again.

On the right, I spotted the concrete slab where Nelson and I had gone fishing. He had called me earlier in the week, leaving a joyously unintelligible message, inviting me over for dinner the night before my ride with the Sabine Pilots. He still had my oil spill fish in his freezer. We cooked them in foil packets on a grill in his front yard, next to his dump truck. The fish that needs no oil, steaming and succulent, with rice and tortillas on the side.

We had reached downtown Port Arthur.

"Isn't this the place where the accident happened in January?" I asked.

"Yes, sir," Tweedel said. "It was a ship just like this." And there was silence on the bridge.

Tweedel and Duane were deeply skilled men, dedicated to their craft and fully aware of its importance. But any system that depends on a high level of human skill is, by its nature, vulnerable to human error. Many months later, when the government finally announced the results of its investigation into the Port Arthur oil spill, it would point the finger largely at the Sabine Pilots. The lead pilot on board the *Eagle Otome*, in particular, had started his turn under the bridge too late, and then failed to correct for the sheering motions that resulted, pushing the tanker into a grand swerve that ended in its collision at the wharf. The government report would acknowledge other contributing factors, but it would place the most specific blame at the feet of the pilots. In the end, it came down to bad driving.

I went outside on the port deck, a steel platform that jutted out from the wheelhouse, high over the water. The rain had stopped, and the breeze was warm under the clouds, and faintly rank. Earlier, the air had been full of birds, a squad of pelicans coasting overhead, just out of reach, and black-headed Bonaparte's gulls cavorting behind the ship. They were attracted to the wake of the vessel, Duane said, to the tidbits churned up from the bottom of the channel as we passed.

And that is how we rolled along. A half-million barrels of oil coasting inland at seven knots, attended by a host of dancing birds. Enough petroleum to sustain the needs of the nation for a whole forty minutes.

\\\\\\\\\\\

A sign once pointed tourists to a viewpoint from which they could peer into Spindletop and see, distantly, the actual site of the Lucas Gusher. But a hurricane blew the sign down, and it has not been replaced. To people driving past, Spindletop is a void space, a low mile of trees by the highway that goes unremarked, even in the area whose prosperity it once sparked.

But however invisible, the wedge of land between Sulphur Drive and West Port Arthur Road holds a secret. And the secret is this: the oil rush on Spindletop is not over. Not quite.

Steven Radley is the last man standing. More than a hundred years and 150 million barrels of oil after Patillo Higgins's hunch first came good—and a half century after the major producers left this land for dead—he is doing his damndest to squeeze every last cup of petroleum out of its stubborn soil.

We met up by a set of large, squat oil tanks that hunkered in the predawn darkness. Radley was a boyish man of fifty, his face creased by decades of work in the oil fields of Southeast Texas. In his truck, we bumped down the dirt tracks that counted as roads on Spindletop, and I asked him about the new well. Was there any chance it would be a gusher?

"I hope not!" he said, smiling. "That'd be fun for about ten minutes. And then we'd have to clean it up."

He was planning to drill to a depth of 1,250 feet, just short of the layer of rock that crowns the salt dome that is Spindletop's dominant geological feature. It was along the edges of this huge underground tower of salt that oil had collected over the ages. The new well would be similar in depth to the famous Lucas No. 1, but unlike its precursor, Radley's well was not about to make oil cheaper than water, or to outproduce the rest of the country.

When I asked him what the new well *would* be, if not a gusher, he grimaced. "Probably not a very good one, honestly."

Beyond the trees there was a cold, artificial glow. We had reached the drilling area, tucked between a woody thicket and a curve in the service road.

For most wells, it is no longer necessary to build a stationary, towering drilling derrick, like you see in old pictures. Instead, Radley was using a mobile rig, mounted on the back of a sky-blue vehicle the size of a fire truck. It had been parked on a level pad laid down on the loose, powdery soil, and then its derrick had been folded upright. A narrow steel scaffold, maybe seventy-five feet tall, the derrick had red and white struts that were fitted with the glare of a dozen fluorescent bulbs.

Radley parked next to the trees. A hopeful thought rattled into my fore-brain. I pointed at the rig.

"You know, I'd be happy to lend a hand."

He laughed.

"No, really," I said. "Just hold a wrench, or whatever."

He laughed again, as though I'd told him some great double–punch line joke about a jerk who wants to help out on a drilling rig—when instead, in a stroke of brilliance, I'd just invented the oil field dude ranch.

We got out of the car. Three roughnecks were clambering around on the derrick. Radley pointed out the driller, the derrick hand, the floor hand. "Just call them roughnecks," he said. "They all do everything." The division of labor broke down on operations this small.

We walked over to the base of the derrick. In front of it, three dozen lengths of drilling pipe, each thirty-two feet long, were laid out. At the base of the rig, lying disconnected on its side, was the rotary drill bit: a trio of knuckled wheels that formed a heavy fist of red-painted metal. Its surface had the hefty gleam of a toolbox.

"That's brand-new," Radley said, nodding with approval. "You can use one like that for four or five wells." After that, it might be possible to rebuild the bit. More likely it would become a paperweight.

Drilling an oil well is an art, one that was developed, in part, right here on Spindletop. The bit is fixed to heavy lengths of drill pipe. The pipe is then turned—driven in this case by a large, hanging tool called a power swivel. The rotating pipe rolls the wheels of the drill bit against the sand and rock below, grinding and shattering downward. At the same time, a slurry of drilling mud is forced down through the pipe, emerging at the bottom of the well from an opening in the bit. As it circulates back to the surface, the mud carries away the cuttings, the loose fragments of rock or sand that the bit has ground through.

"Drilling mud was invented right here!" Radley said, as we watched the roughnecks prepare the rig. "They actually had a bunch of cows tramping around in a pen to produce it." It was Anthony Lucas who had pioneered the use of drilling mud, and it was a key innovation. Rotary drilling had been around for a while by the turn of the twentieth century, but the use of drilling mud prevented the narrow sides of a well from falling in against the shaft of the drilling pipe, causing it to seize up. This was especially critical in the young geology of coastal Texas, which confronted drillers with layer after layer of clay and sand. Drilling mud has been a staple of oil exploration ever since, even on the most sophisticated offshore rigs.

Radley's roughnecks threw themselves into their preparations like lively, oil-stained pirates. The bit was fitted to the heavy first stage of pipe, and the driller mounted the control station. The power swivel swiveled. Mud circulated. A mercenary focus concentrated the air; I stared at the rig like a midshipman watching the sails for wind. The sun had come up. Then, at the pull of a lever, the rig's engine revved, the shaft spun, and the bit dropped through a hole in the drilling floor into the ground below. Returning mud flowed up out of the well and along a small trench to a large pit out back, where it would be filtered and cycled back into the pipe. We were drilling for oil.

Radley's gear was recent technology: powerful, mobile, and automated. Tasks that even a few decades ago required multiple workers—like adding a new section of drilling pipe—could now be accomplished by two people with a power swivel and a pair of hydraulic tongs. But the basic elements of rotary drilling—derrick, pipe, bit, mud—hadn't changed. Anthony Lucas would have known with a single look what Steven Radley was doing, and why. The difference is that, while Lucas was convinced he could find large reservoirs of oil, Radley wasn't looking for anything but dregs.

Most oilmen focus on discovering and extracting deposits that haven't yet been found or tapped into. From the speculators buying up land around Beaumont in 1901 to multinational corporations using sophisticated seismic imaging off the coast of Brazil, the purpose of the game is to secure a prize that can justify the vast up-front investment.

Eventually, though, the returns begin to diminish, and the expense of reconfiguring and maintaining existing wells—not to mention sinking new ones—is no longer justified by the declining revenue coming out of the field. It's at that point that an oil company abandons a place. Not when the oil is completely gone, but when there's too little of it to be worth the effort.

Such a field becomes known as a *stripper* field, and will be left for smaller companies with less overhead or less ambitious profit targets. Such companies depend on keeping their costs down and on using whatever technology or ingenuity they can to squeeze out of the ground what Big Oil left behind.

"I don't have a lot of overhead," Radley told me. "We do things the least expensive way possible. I've got it down to where I don't think anyone can drill cheaper than we do."

He started drawing lines in the sand, counting off the strikes against big oil companies. They had CEOs, and executives, and lawyers on staff, and on and on. In contrast, Radley's corporate structure was uncomplicated: He owned 40 percent of the company, and his father owned the rest. And unlike the chief executives and board members of, say, ExxonMobil, Radley and his dad were their own middle management and technical staff, working their own leases every day. Radley's wife ran the office, and his son worked for the company as well. (The roughnecks were freelancers.)

Radley's other trick was to avoid rental and contractor fees by owning his own oil rig. Usually, a small operation like this would contract the drilling to someone else. But not Radley.

"That's mine!" he said, pointing at the drilling rig. "That bulldozer, that's mine! Everything out here is mine! That's why I can make it on a five-barrel well. And we do our own geology."

This meant they saved money by forgoing sophisticated geophysical analysis, like seismic reflection or gravity surveys. Instead, they used a simpler method. Radley demonstrated it for me by scratching his head and then pointing at a spot on the ground.

"This looks good!" he said.

By eleven in the morning, we were 250 feet deep and settling into the rhythm of the work. The rig's engine would rev up, and the section of drilling pipe would descend into the ground, sometimes fast, sometimes slow, depending on the character of sand or rock the bit was working through. Once the section of pipe had fully descended, it was time to detach the power swivel, pull another piece of heavy pipe off the rack, hoist it upright, and thread it into the string. One of the roughnecks would reattach the power swivel to the top of it all, and then drilling would begin again.

It wasn't easy work. The roughnecks were in constant motion, guiding the new lengths of pipe into place, making sure they sat correctly, spraying excess mud off the drilling platform with a hose, readying the next stage of pipe. Radley timed the intervals between stages of drilling, to see how efficient his workers were in the changeover. "About three minutes," he said. "Pretty good."

This was Radley in the role of both "pusher" and "company man," the two people whose job on a well is to make sure that it gets drilled without wasting time or money. And if Radley was a bit casual as a pusher, that was only because he knew and trusted his crew. The most important thing, in his view, was that he manage the well in person.

"On those big rigs," he said, "the pusher's in a trailer, he's got screens with pressure, drill speed, temperature, how many feet per minute, per hour, or whatever. And he just sits there and watches the screens."

He shook his head. "To me, that's not drilling. That's *bullshit*."

Section after section of the drill string descended into the ground, and we got bored. Radley and I hopped into his truck and took a spin around the lease. He had owned oil wells in one form or another since he was fifteen years old, when his father had encouraged him to invest in one. His father had been working his own small-time wells on the weekends and after hours from his job as an electrical contractor. Radley had started going along to help when he was still a little boy. It was all part of the world of Little Oil, in which it was completely reasonable for a teenager to buy his own share in a well, and where a mom-and-pop company could end up with the mineral rights to the oil field that sparked the petroleum revolution.

The problem with sinking wells is that it's hard to tell which will produce and which won't. We passed a nearby well that wasn't producing anything; eighty feet farther was one that yielded four or five barrels of oil a day—and fifty of water. The small, lazy rocking horse of the pump dipped up and down as we drove by. Radley told me that just on the other side there was a well that produced even less oil, but more water. The area was all fractured, he explained. The ground was still shifting. It was impossible to know exactly where the oil was, in what direction it wanted to seep, and where the blockages were. Even Lucas No. 1 would have come up dry if it had been drilled fifty feet away.

"Oh, I would love to go down one of them holes," Radley said. "We can't know what's down there." It was only through deduction that he could become anything less than blind, piecing together an idea of what was going on at the business end of his drilling pipe, hundreds or thousands of feet underground.

We passed another well, and another. Radley estimated that he had thirty-five on the two leases he operated on Spindletop. Added to what he produced from a few other leases he owned, his company pumped about a hundred barrels of oil a day. I couldn't decide whether that sounded like a lot. Was he making much money?

"I don't even watch the price of oil anymore. I haven't looked in three weeks," he said. But of course he couldn't ignore it completely. "At forty dollars a barrel, it'd be eating on me. When it was a hundred, hell yeah, I was happy. It's a living. I ain't gonna get rich." He chuckled. "But I'm gonna eat real good."

We headed back to the drill site, passing a wide pond. "You can fish in that water," he said. "There's bass. But they're wormy."

For a long time, Radley had had the area almost completely to himself. "It used to be we were the only ones out here," he said. "It was literally just me, my dad, and my son." But life on Spindletop had changed in recent years. There was a new wave of activity, and now perhaps a hundred people were working there on any given day. The natural gas business had landed.

But it wasn't here to extract the stuff.

In a weird twist, the real action on Spindletop was no longer in taking fossil fuel out of the ground but in putting it back. For a range of economic and logistical reasons, natural gas companies need to store their product in large quantities, and creating caverns in underground salt formations is one of the best ways to do it. The huge salt dome underneath Spindletop, along the flank of which so much oil once collected, was now being hollowed out with caverns, each several hundred feet in diameter, thousands of feet tall, and more than half a mile underground. Gas storage companies planned to use them as impermeable storage tanks, each of which could hold billions of cubic feet of natural gas.

So the hill was a giant layer cake, divided for different purposes according to depth. The storage companies took the bottom layer, from about 6,000 to 2,500 feet belowground. From 1,500 on up was Radley's territory, the specific depth for which he held the lease for extracting oil. Then, on top of it all, there was one last, tenuous layer.

"You see those orange fences over there?" Radley said when we were back at the drilling site. "Those are archaeologist playpens."

The top several inches of soil on Spindletop were also being pros-
pected—for historical evidence. A university archaeologist had fenced off
the areas he thought most promising for his investigation of the early oil
industry.

"It's kinda neat being part of history," Radley admitted. "The world
changed right here." But he took a dim view of the archaeologist, who Radley said hadn't been coming to the site nearly often enough to get his work
done—and get out. Meanwhile, Radley was supposed to stay out of the areas
enclosed by the orange fences.

"I've got to get in there and drill a well," he said, plainly annoyed. "I'm
trying to be patient." It was a testy relationship. Radley had already intruded
once into a fenced-off area to do some maintenance on an electrical line.
The archaeologist had complained, upset that the area had been disturbed.

Radley shook his head, lips pursed. "I said to him, 'I been in there with
a dozer ten years ago. It's already *been* disturbed.'"

In the early afternoon, drilling on the new well broke down. The power
swivel was leaking mud, I think, and needed to be pulled apart. Radley's
father, a cheerful man in his mid-seventies who looked likely to show up to
work for at least another twenty years, inspected the power swivel, which
now lay on the drilling deck, ready for him to operate. "This business would
be okay," he muttered, "if it wasn't for breakdowns."

I walked over to one of the orange fences set up by the archaeologist.
KEEP OUT! a sign read. AVOID ALL CONTACT AND ENTRY TO THIS AREA. I
peered over it, into the history of Spindletop. Tall grass grew out of the
gravel.

The last section of drill pipe was just being pulled out of Radley's new
well when I showed up again three days later. The drill head came out caked
with earth, dripping with drilling mud, fresh from its journey a thousand
feet into the earth.

"We didn't get to what we wanted," Radley said. He had called off the drilling around 1,150 feet, 100 feet short of their goal. "We were drilling that gypsum, we think it was, and it's just so slow. A foot an hour, or slower." So they had stopped.

I wished Radley a happy Earth Day. "Is that what it is? Well, we got some earth right here," he said, pointing at the drilling rig.

The question was what kind of earth it was. Radley and his crew were waiting for a contractor to log the well, lowering sensors to measure the properties of the soil and rock at different depths, and determine whether it was likely to produce oil. (Logging his wells was one of the few things Radley couldn't afford to do himself.) This was the critical moment, on the basis of which Radley would decide whether to make the investment of lining the well and outfitting it with a pump or to cut his losses and go drill his next well.

While the roughnecks horsed around and told jokes—it was the first time I had seen them at ease—Radley and I leaned on the bed of his truck and waited for the logging to start, and soon we had begun the political debate that I had known was coming ever since we first met. In a matter of minutes Radley was pounding his fist on the truck and telling me in a full shout that Obama was "not an American." I won't even repeat the things he said about people in Africa, and how they were responsible for the world's overpopulation. I told him that where I came from, people went to hell for talking like that. He leaned on the truck and let out a giant sigh.

With that out of the way, I asked him what was in store for the oil industry. Wouldn't supplies dwindle someday? Would his grandchildren be able to spend their lives on an oil field?

"Oil is actually a renewable resource," he joked. "Just not in our time. There's still shit down there rotting and decaying. It just hasn't turned into oil yet." But even in the short term, he wasn't worried about oil running out. "It may get scarce," he said, "but it won't run out. The technology isn't there to reach a lot of what's there, but it will improve. There are big pockets of oil offshore that ecologists won't let us drill. I love ecologists. They keep the price of oil up."

But that brought us to the question of whether the remaining oil *should* be drilled. I asked him what he thought of climate change.

He let out a deep breath. "I think what we're seeing is just the Earth's natural cycle," he said, and kicked the powdery soil by the truck. Human emissions might have some effect, he allowed. "But not as much as people say. Al Gore, he's full of shit.

"Look," he said, cutting to the chase. "If you drive an electric car, you still have to get the electricity. What makes electricity? Oil. Coal. And who made the tires? Who made the plastic? Who transported it to you? Everything in this world is affected by oil."

It was the same chase that everyone else cuts to. Whether they're celebrating the fossil fuel economy or execrating it, everyone genuflects to oil's market-finding, world-powering genius. In both cases, there's an undercurrent of fatalism in the cataloging of oil's uses, a recognition of how difficult it would be simply to unmake the choice of fossil fuels as a basis for our society. The investment is so total—in infrastructure, in industry, in our way of living—that oil cannot simply be swapped out for another source of energy or materials, however much promise the alternatives may hold. Until another industry actively displaces its uses—or until scarcity makes it impractically costly—oil will not simply abandon the markets it dominates. Nor will the uncountable people and companies that make up the universe called *the oil industry* simply give it up. Not while it still has life in it.

Maybe, then, Spindletop is not only a relic of oil's past but also a vision of its future, however distant. Maybe one day this is what it will come to: every oil field a field of stripper wells, managed by a single family. A lone oilman, with his own derrick and his own bulldozer, producing locavore petroleum for refineries down the road. That's the power of the long-since-taken path. You don't unmake such a choice. You ride it into the ground.

\\\\\\\\\\\

On my way out, I visited Lucas No. 1.

Earlier, Radley had pointed out the spot, down a short gravel road that dead-ended on a shallow pond. A flagpole stood on the shore. But there was no flag. Just a lonely exclamation mark of metal planted in a squat trapezoid of concrete. "That's the one that done it," Radley had said.

It had been raining, and we didn't stay long. But today was windy and bright, and I had the run of Spindletop. I drove along the dirt service roads until I found the turn.

It was a peaceful spot, the only noise the blind pinging of the flagless, wind-driven rope against the flagpole. I ran my hand over the rough concrete of the base. On one side, it wore a metal badge embossed with the tiny image of a derrick fountaining oil to twice its height. I leaned over to read the words engraved on the medallion.

SPINDLETOP GUSHER—LUCAS NO. I—ORIGINAL LOCATION

I scrambled onto the top of the base and hooked my arm around the flagpole, looking out at the marshy lake. It was streaked with some kind of algae or floating weed, pushed by the wind into clumps on the near shore. The distant sound of a train floated on the wind, the clanging of tanker cars being jolted together. Far off to the right, I could see the pump on one of Radley's wells.

Two days earlier, about three hundred miles up the coast, an offshore drilling rig had exploded. Now it had sunk. In one of the country's worst environmental disasters ever, an open underwater well was giving out fifty thousand barrels or more into the Gulf of Mexico every day. There were still real gushers.

Soon, every piece of containment boom in the country would be in Louisiana. Armies of workers would arrive, flocks of media, the National Guard, the president. Oil Mop's boats would be starting their engines. Rhonda, the grumpy pelican lady, would soon become the wildlife director of BP's oil spill response. Another disastrous bloom, and thousands flocking to its spectacle and wealth.

The *Deepwater Horizon* spill would dwarf anything that had ever happened in Port Arthur's ship channel. But as a gusher, it wouldn't touch Lucas No. 1, which had thrown as many as a hundred thousand barrels of oil into the sky in a single day—on this very spot.

I stood on the concrete cenotaph, next to the flagpole, and thought of the water gusher at the Spindletop–Gladys City Boomtown Museum. I had gone there again this morning, with a hundred dollars, which I had given to the nice lady in the gift shop—the gift shop that sold souvenir vials of Spindletop crude, provided by Steven Radley. A man called Frank, who knew how to turn the spigot on, had come by—they called him the Gusher Guru.

The replica derrick stood in front of us, in the broad field outside the museum, where the obelisk marks the wrong spot. The Gusher Guru was in his late eighties. In a thick drawl, he told me he had worked all his life in refineries and on oil wells. Now he did this.

"You ready?" he asked.

"Yup," I said.

He pushed a green button on the exterior wall of the gift shop, and we turned to look at the derrick. There was a hiss, a gurgle, and then water erupted out of the nozzle, brown at first, then white—*white*—oil's perfect inverse, because they do not mix, roaring and explosive.

I walked over to the derrick. Inside, water was blasting out of the nozzle, a sparkling, violent froth. I looked up along its silver length. It crashed upward, battering the interior of the derrick as it burst out the top and hurled itself into the air. I walked to the other side, to where the water was falling, and it drenched me, cold and clean, pelting me with clear pebbles that glittered in the sun.

Now, at Lucas No. 1, I tried to imagine the violence of that water gusher springing out of the flagpole's base, but thick and black and green. I stared upward, at the space into which the Lucas Gusher had exploded. But I couldn't quite see it. The more I tried to picture it, the more I felt its absence—an empty blue volume where oil had said its name.

The sun blinked. A hawk had crossed it. The shadow coasted away on the ruffled surface of the lake. It was Earth Day. Half a mile back, Steven Radley was sinking well after well into the miserly ground. In the Gulf, BP's well was running and running. It would be months before anyone could stop it.

But not Radley's well. He called me before I left town. The well was dry, he said. They were going to plug it and move on.

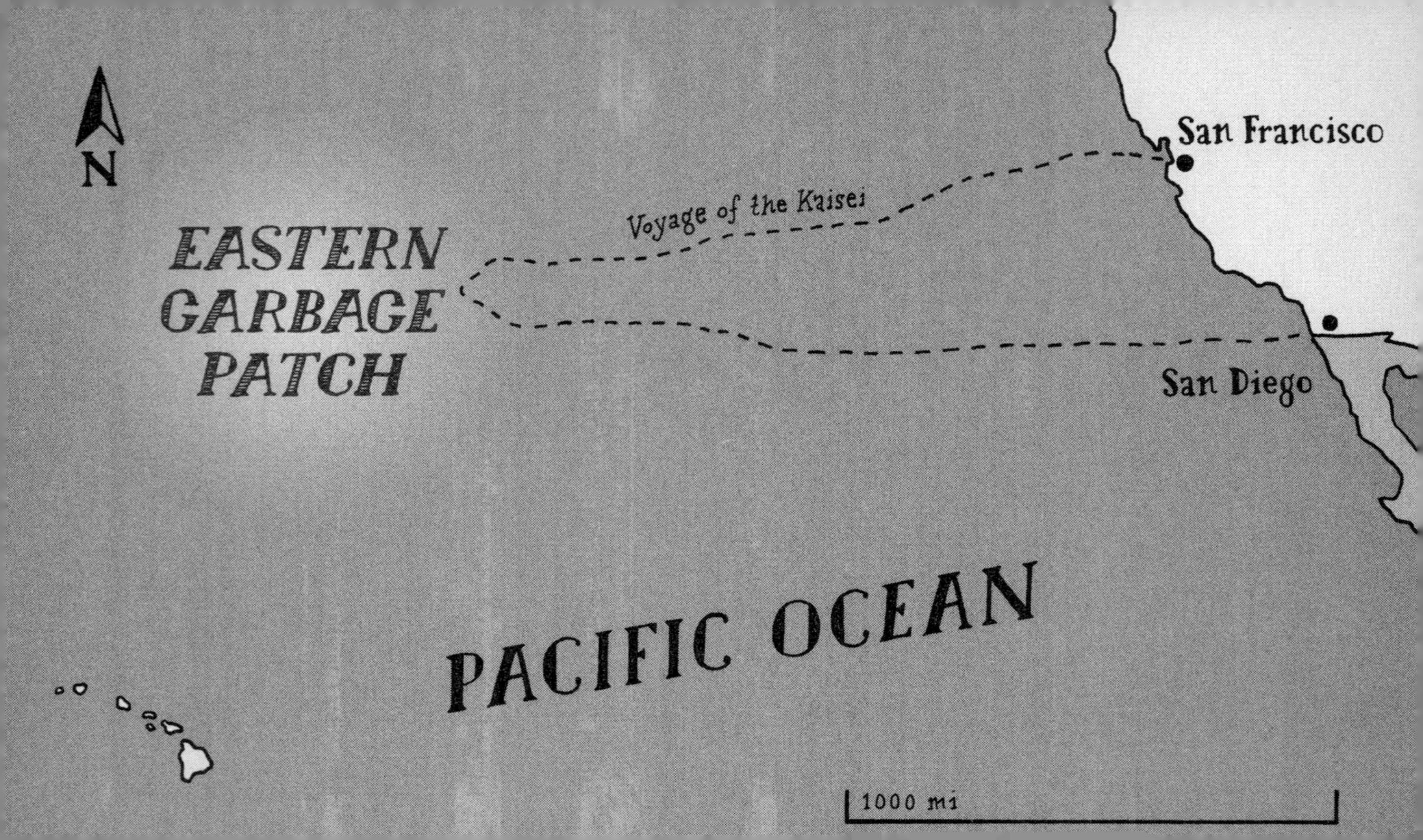
N
EASTERN
GARBAGE
PATCH
Voyage of the Kaisei
San Francisco
San Diego
PACIFIC OCEAN
1000 mi

FOUR

THE EIGHTH CONTINENT

In my sleep, I heard the call. *All hands.* Someone had shouted it into our cabin. *All hands to strike sail.* We fell out of our bunks, struggled into our rain gear, and went above half-awake.

The deck was a starless uproar of wind and sound. "The Navy's running an exercise nearby," said the first mate. "They've ordered us to head north. I asked them to let us run downwind, but they just repeated the order." The ship, under engine power, was running directly into the wind, the sails flapping powerless and wild. They would be torn to shreds.

We wrestled the foresails in the dark. The air filled with spray, with thick rope jerking and snapping in chaos. There were six of us in the bow. Four were out on the bowsprit—the long spar extending forward over the water—and two on the most forward part of the deck, where the bowsprit joined the ship.

I was on the deck, feeling with my hands in the dark, trying to find the downhaul lines and gaskets that would draw down and fasten the sails. After weeks at sea, I knew what I was looking for, but that didn't mean I could find it.

The ship crested a large wave. We felt the bow rise higher and higher into the night. It seemed to pause at the top. For a moment we floated in the salty air.

Then we fell. The ship buried its prow in the oncoming wave, deeper than ever before. The four on the bowsprit—my friends—disappeared below the surface, foam churning over their heads. Were they clipped in? The deck went under with them. The water surged to my waist, tugging at me, sliding me aft. Robin grabbed my arm and I grabbed the rail, and we kept ourselves from tumbling backward down the deck. I looked at the bowsprit and thought, *All I see is foam.*

A second more and the ship came through, rising out of the swell, and I saw them. They were still there, still clutching the bowsprit, all four of them. I counted them again. *Four.* Had there been more?

"Is everybody there?" I shouted. "Is everyone still on board?"

But they were already working again, grappling the sails, water streaming from their jackets, shrieking like bull riders.

Robin let go and we returned to the tangle of lines at our feet. But my head was swimming with the afterimage of the water rising up to us, of the sea invading the deck. I still felt it, how it pulled at my body, an overwhelming force that swirled around and through us, the alien gravity of another universe, the black remorseless ocean.

\\\\\\\\\\\\

You will have heard of the Great Pacific Garbage Patch: an island of trash, formed by a giant vortex of currents that gathers all the eternal, floating plastic in the northern half of the Pacific Ocean into an endless, swirling purgatory, a self-assembling plastic continent twice the size of Texas.

Let's nip this in the bud: It's not an island.

I'd like to say that again. It's not. An island.

There is no solid mass, no floating carpet of trash, no landfill. But it *is* real. It was first discovered in 1997 by the yachtsman and environmentalist Charles Moore, who made it the focus of his nonprofit, the Algalita Marine Research Foundation. It is thanks to Moore's observations that the Pacific Garbage Patch entered the popular consciousness, sometime in the mid-

2000s. As for who's responsible for the irresistible image of a plastic island, I don't know. But someone should run them down and give them a nice, quick smack. Furthermore, an exorbitant fine should be levied on anyone— *anyone*—who describes this non-island as being "the size of Texas" or "twice the size of Texas." When I was doing my preliminary research, it seemed impossible to find a piece of media about the garbage patch that *didn't* mention Texas.

Why Texas? Is there no other territory that could serve as a reader-friendly reference point? Has hack journalism become so impoverished an art form that its practitioners can't even be troubled with the five googling seconds it would take to craft an entirely original gem like "three times the size of California," or "two Nevadas and an Arizona," or "nearly as big as Alaska, if you leave out the Aleutians"?

The real problem is that, although two Texases clear a trim half-million square miles, nobody knows how large the Garbage Patch actually is. Unlike Texas and, critically, *unlike an island*, it has no defined boundary, only a general area. So let's just call it big, and be done with it.

A more appropriate analogy would be that of an ecosystem. *System* is the key here, implying something much more complex than a simple floating object. From tuna-size hunks of Styrofoam and discarded fishing nets that lurk like massive jellyfish, down to microscopic pellets that hang in the water like artificial plankton, it is a vast, plastic simulacrum of the living ocean that is its host. And precisely because it is so complex, and so far from land, its nature is poorly understood.

Nobody can say for sure exactly where all the stuff comes from, but there is broad agreement that its sources are disproportionately land-based. A surprising amount of trash manages to avoid the landfill, and when it does, it often makes its way to the sea, whether by way of storm drains, rivers, or other avenues.

Since plastic objects don't degrade easily, if ever, they have plenty of time to work their way out from land and find the ocean's currents. A plastic bottle taken by the currents off San Francisco will travel south as it heads out into the Pacific, passing through the latitudes of Mexico and even Guatemala before heading west in earnest, caught by the North Pacific Subtropical Gyre. This vast counterclockwise vortex will take the bottle clear across to the Philippines before shooting it north toward Taiwan, close by Japan, and then spitting it back past Alaska, toward the rest of North America.

Around and around the Gyre goes, and the plastic bottles and hard hats of the North Pacific go with it, we assume, until at last they drift into the becalmed zones spinning at the eastern and western ends of this oceanic conveyor belt. These are the Eastern and Western Garbage Patches. (The Eastern gets all the attention because it's closer to the United States and was the first to be discovered.) Here, our plastic bottle finds its friends: all the other bits and pieces of plastic that have made it into the ocean in the previous who-knows-how-long. And here they wait, year upon year, breaking into fragments from the action of the waves, and strangling hapless turtles, choking overzealous albatross that mistake plastic for food, and being eaten by fish.

Eventually, the scientists and the activists and the adventurers come. Whatever part of our plastic history that floats, the Garbage Patch is the place for them to find it: our bottles, our plastic tarps, our popped bubble wrap, the tiny plastic "scrubbing beads" of our exfoliating face soap. It's all here for the hunting.

Or so I hoped. But without a single cruise line running through, how was I to know for sure? Which brings us to another interesting thing about the Garbage Patch: hardly anyone has actually seen it. It takes serious oceangoing chops to get out there. And there's almost no reason anyone with a boat would bother. Most people with yachts and things are more interested in going to places like Hawaii, or the Bahamas, or *anywhere*. But the Garbage Patch, inherent to its formation, is in the middle of the biggest nowhere on the planet.

The Gyre had seen several expeditions from researchers and activists the previous summer, in 2009, so I took to the phones, beginning a campaign of sustained pestering that I hoped would be my ticket onto one of this year's voyages. And that's how I met Project Kaisei.

\\\\\\\\\\

I found the *Kaisei* docked in Point Richmond, across the bay from San Francisco. A steel-hulled, square-rigged, 150-foot-long brigantine, it was a striking sight. Think *metal pirate ship* and you will have the image. The ship is the namesake and floating linchpin of Project Kaisei, a nonprofit venture dedicated, as its motto reads, to "Capturing the plastic vortex." I had somehow convinced Mary Crowley, one of its founders, to let me come along on a three-week voyage to that plastic vortex, a thousand miles away, but I had my doubts about capturing it.

Especially if we never left. We had spent more than a week without a clear sense of when we might set out to sea. A departure day would be announced, only to dawn with the new radar unit still absent, or with provisions yet to be delivered, or with a cook not yet hired, and we would not sail.

In the meantime, a subset of the crew would show up each day to help clean the boat, patch its rust holes, touch up the blue paint on the hull, or install an extra life raft, and I had time to develop my mixed feelings about the *Kaisei*. From the moment I first stepped aboard, I had tasted that flavor of excitement that has a note of terror. She had two great masts, the forward one boasting four spars: the yards, from which majestic square sails would drape, sails that belonged in a biography of Lord Nelson. Dozen upon dozen of cables and ropes—*lines*, we learned, not ropes but *lines*—led from wooden pins on the deck to points above; this set of lines to pull a sail down, that to pull it up; lines to orient the yards to starboard or to port; lines to raise and lower the spar of the gaff sail; lines to raise and lower sets of pulleys that were connected to still further lines.

Was I going to be asked to climb those masts, to edge out along those yards, approximately a thousand feet up? Like most sensible people, I don't really have a fear of heights—only a fear of falling to my death. Which is not a fear at all, but a sensible attitude. On the other hand, what is the point of being on a tall ship if you don't experience the tallness? I knew that when asked to go aloft, I would overcome or at least bypass my fear and force myself to do it. And so what I really feared was that I wasn't afraid enough.

This was all neatly analogous to my broader situation: instead of a nice, short jaunt on a press boat or a proper research vessel, I was going to sea for three weeks or more. A thousand miles from land when I wanted to be at home in New York, when I *should* have been at home, squaring away wedding plans, preparing for the moment of my good fortune, only two months away, when the Doctor and I would get hitched. And the *Kaisei* would be sailing in total seaborne isolation. There would be no satellite phone for the crew, no data connection, no way to communicate with my family or with the Doctor. No way even to apologize, once I went, for being gone.

The ship itself was charming, if a bit scruffy, with cabins that were cozy but not claustrophobic, and a pair of lounges ample for a small crew, and decks of faded wood. In front of the wheelhouse, with its radio and its radar display, was an outdoor bridge, where the deck rose into a platform facing a large, spoked wheel. It was the kind of wheel I would have expected to see on the wall of a nautical-themed restaurant.

The problem was not the *Kaisei*. The problem was us. As the days went by, spent in sanding and painting and offloading unneeded scientific equipment from the previous year's voyage, I met the volunteers who would be the crew. How many people did it take to sail a 150-foot brigantine? I wasn't sure we yet had ten. And as we got to know one another, it emerged that very few of us knew anything that would be useful in the safe operation of said brigantine.

There was Kaniela, for instance, an affable young surfer from Hawaii and one of the hardest workers on the boat. He asked me if I knew much about sailing.

I didn't, I said. Not a thing. You?

Nah, man. I'm hoping to learn.

Then there were Gabe and Henry, two recently graduated Oberlin hipsters. The morning we met, they were standing on deck huddled against the early chill, hands stuffed in their pockets, wearing their sunglasses. A surly pair, I thought, but they turned out just to be badly hung over, and had brightened up by mid-afternoon. They told me they both had degrees, more or less, in environmental studies, or something. Upon moving back to Marin County from Oberlin, they had gotten internships at the Ocean Voyages Institute, the umbrella organization for Project Kaisei. But three weeks at sea seemed a little extreme for an internship. I asked them why they were coming.

With a straight face, Gabe told me that he was here for the adventure. He wanted to be an adventurer. A *rakish rogue*, he specified. And this was the first step toward his goal.

The ravings of a contaminated mind. I turned to Henry. I asked him if either of them knew how to sail.

He smiled. It was a thin smile, similar to a wince. They had taken sailing in high school, he said. Little two-person boats.

What was that feeling in my gullet? Desperation? I made my way from volunteer to volunteer, making a mental map of our skill set. We had a deep bench in watersports and the teaching of high school science. Otherwise, it was a mixed bag. There was a boatbuilder, a former journalist, a few students. They were all interesting, thoughtful, hardworking people who didn't know a damn thing about sailing a tall ship.

I put my hope in the second mate, a calm, confident tall-ship sailor . . . who quit. After a single afternoon on board, he told the captain he didn't like the look of things and got the hell out of there.

There it was again. That sinking feeling.

The votes of ill-confidence started to pile up. A team of Coast Guard bluesuits came to inspect the boat's papers. As they left, chuckling, I heard the *Kaisei*'s captain say, "They'd never seen anything like this."

The more I learned about the *Kaisei*, the more I realized that, from a technical point of view, she was an oddity. One evening I sat on the aft deck with the ship's engineer, watching Richmond's tugboats go by and listening to him complain. The engineer was probably the most important person in the crew, if it mattered to us that the ship remain afloat, that we have fresh water, and that the navigational gear function. Night after night, he had been up late, coaxing the ship's systems into fighting shape. He was grumpy, but that seemed like a good sign. You don't want a laissez-faire engineer.

The *Kaisei*, he told me with some exasperation, had been built in Poland, only to be refitted and operated in Japan. Everything was in Polish and Japanese. And the electricity. He shook his head. Multiple standards, in a dazzling range of voltages. The irregularity extended literally to the nuts and bolts of the ship: some were metric, some weren't, and so multiple sets of tools were required, though none of the multiple sets on board were complete.

The engineer sipped from his mug and let out a great sigh. "Excuse me," he said. "I'm into my cup."

Within a few days, he, too, had quit.

We now had no second mate, and no engineer, and none of us lowly volunteers—the crew—knew what the hell was going on. Every day of delay was shortening the mission: in barely three weeks, the *Kaisei* was booked to participate in the San Diego Festival of Sail, where we would blow the minds of all those tall-ship enthusiasts with our adventures in deep ocean plastics. So every day tied up at the dock was a day we didn't spend in the Gyre. We began to doubt that the boat would ever leave the dock. And with experienced crew members disappearing by the day, the rank and file were wondering if we, too, should step off the boat.

Something held us back, though. Something that counterbalanced all the bad omens. A single factor that kept the entire crew from walking.

It was the Pirate King. His name was Stephen, and his position was first mate, but I thought of him as the Pirate King of the *Kaisei*, a single person so compulsively knowledgeable about seafaring that he made up for the frightening deficits in the rest of us. A compact man, even short, he was trim and strong, with a close beard and two golden hoops in his left ear, and just in case we weren't paying attention, he wore a black baseball cap decorated with a skull and crossbones.

The *Kaisei* had a captain, but we mostly ignored him in deference to the Pirate King, who exemplified that very specific kind of manhood that is built on overwhelming knowledge. He knew how to navigate, how to tie knots, how to rig a sailboat, how to walk along the yards with barely a hand to hold himself in place, and how to slide down the stays, Douglas Fairbanks–like, landing himself back on deck in mere seconds. He never wore a safety harness. He knew how to furrow his brow and raise his voice and tell us that, as first mate, he was responsible for us. He had personally circumnavigated the globe in his own small sailboat at near-racing pace, sailing through every kind of sea imaginable, even surviving a rogue wave. He was equal parts Jack Sparrow and Han Solo, and we would have followed him anywhere—across the Pacific in a rowboat, up Everest in our shorts, suitless through an airlock into deep space—as long as he was there to tell us what to do. You can actually survive in deep space without a space suit, he would have explained. You just need to control your exhalations.

He promised us he would walk off the boat if it wasn't safe, and that was good enough for us. He became our knot-tying, aggressively all-knowing weathervane. And sure enough, a new engineer was found, and a cook, and everything came together at the last minute, and finally, eventually, suddenly—we sailed.

The crew of a ship about to go out of range becomes diligent with its telephones. I texted my friends and family and posted a picture from the far side of the Golden Gate Bridge. I received one, too: my friend Victoria had gone up on the Marin Headlands to take a picture as we left. On the boat, we looked at it, the picture of ourselves. It showed the mouth of the bay, opening out from the land of the Golden Gate. Our ship was in the center of the picture, our huge steel ship, barely a dozen pixels wide, the merest smudge against the sky-colored sea.

And I talked to the Doctor one last time. She made me promise her something. She made me promise that if I found myself being washed off the ship, I would hang on.

"Promise," she said. "Promise you'll hang on with your last fingernail."

\\\\\\\\\

Conversations about the Great Pacific Garbage Patch tend to follow a certain profile. First there is the flash of recognition, embedded with a nugget of misinformation:

Right! The giant plastic island! The one the size of Texas!

It's not an island, you say.

Well, right, they say, moderating. *It's more of a pile.*

You narrow your eyes. Seriously, how do you *pile* anything on the ocean?

Eventually, with coaxing, they let go of the island imagery, of impractical notions of how things pile, of Texas. Sobriety achieved, there comes the inevitable question:

Can it be cleaned up?

A lot of people have considered this question, and have debated it, and have pondered different strategies and possibilities. From this, a broad consensus has emerged among scientists and environmentalists, which I'm happy to summarize:

Get real.

We're talking about the ocean here. Even assuming that we could just get a big net—whoever *we* is—and that it would be worth the massive use of fuel to drag it back and forth for thousands of miles across the Gyre, and that there would be an exit strategy for what to do with a hemisphere-size net full of trash . . . even granted all these impossibilities, there remains the intractable fact of the confetti.

As a plastic object spends year after year in the water, it becomes brittle from the sun. The waves begin to break it into pieces, and gradually it is delivered into smaller and smaller bits, a plastic confetti that might be the most troublesome thing about the Garbage Patch. Nets and larger objects may strangle marine life, and bottle caps and disposable cutlery may fill the stomachs of baby albatross, but the confetti has a chance of interacting with the ecosystem at a more fundamental level. Since it is consumed just as food would be, it has the potential to introduce toxins at the bottom of the food chain, toxins that may be concentrated by their passage up the chain to large animals like tuna and humans. In 2009, researchers from the Scripps Institution of Oceanography (on a voyage funded in part by Project Kaisei) found plastic in the stomachs of nearly a tenth of all the fish they sampled in the Garbage Patch, and they estimated that tens of thousands of tons of plastic are consumed by fish there every year.

This is a lot like what happened in Chernobyl, where radionuclides followed the same pathways as nutrients to become incorporated into vegetation, and presumably animals. As the journalist and author Mary Mycio has written, in Chernobyl "radiation is no longer 'on' the zone, but 'of' the zone." Perhaps we can already say the same of plastic in the oceans. It is not only a fact of life, but part of it.

How then to clean it up? To remove a billion large and tiny pieces of the ocean from itself? A cosmic coffee filter? And then, how to avoid also straining out every whale and minnow in the sea, every sprite of plankton?

It was no surprise, then, to find that organizations devoted to this issue tended to avoid the idea of cleanup. Charles Moore's Algalita Marine Research Foundation, a leader in the budding field of Garbage Patch studies, has a bent for "citizen science" that hearkens back to science's roots as a discipline founded by amateurs. Rather than speculate about cleanup, it produces peer-reviewed research for journals like the *Marine Pollution Bulletin*. Moore has been openly scornful of the idea of cleaning up marine plastic. Appearing on the *Late Show with David Letterman*, he swatted down his host's hopeful questions about cleanup. "Snowball's chance in hell," he said. (Letterman told Moore his outlook seemed "bleak" and proposed they get a drink.) Other organizations focus on finding garbage patches in the other ocean gyres of the world or on raising awareness to combat the overuse of plastic on land.

So Project Kaisei is special. "Capturing the plastic vortex" is more than its motto. It's a succinct mission statement. Not content to tilt at the windmill of keeping plastics out of the ocean in the first place, Kaisei has chosen to go after the biggest windmill of them all: finding some way to clean them up.

The force behind Project Kaisei is Mary Crowley, a toothy woman in late middle age with a warm smile and an unshakable belief in the possibilities of marine debris cleanup. She has gone so far as to envision ocean-borne plastic retrieval as an actual industry. "Fishing for plastics, so to speak, is not that different from fishing for fish," she told me, leaning on the *Kaisei*'s starboard rail. "And unfortunately, we've so overfished the oceans. I think it would be a wonderful employment for fishermen to be able to get involved in ocean cleanup, and give fish a chance to have a healthier environment and restock."

We can only hope that one day the fishing industry will be rescued by fishing for plastic instead of actual fish. (Indeed, a proposal to subsidize fishermen for debris pickup has even been floated in the EU.) In any case, let's state for the record that in the early-twenty-first century, when most people said cleanup was impossible, Project Kaisei kept the dream alive. May they be proven prescient.

This summer's mission, though, had been shrinking in scope almost since it was conceived. There had originally been plans for *two* trips, in quick succession, as well as a short press voyage that would depart from Hawaii to rendezvous with the *Kaisei* in the Garbage Patch. Mary had even spoken to me of tugging a barge out to the Gyre, of recruiting fishing boats to help retrieve mass quantities of refuse.

Those ideas had evaporated, and the scope of the mission had narrowed. The goal of the voyage now, Mary told me, was to use ocean-current models being developed by scientists at the National Oceanographic and Atmospheric Administration (NOAA) and at the University of Hawaii to pinpoint the areas with the largest plastic accumulation. By comparing our observations with the scientists' models, it would be possible to devise effective ways of finding the plastic, a critical precondition for future cleanup. Think of it as fisheries research for the seaborne trash collectors of tomorrow.

She said we would also be "working on the most effective ways to use commercial ships—tugs, barges, fishing boats—to do actual collection," or, as a Project Kaisei press release put it, "further testing collection technologies to remove the variety of plastic debris from the ocean."

The word *further* alludes to the *Kaisei*'s voyage of the previous summer. I heard many references to the technology developed as part of that voyage, specifically "the Beach," a device designed to answer the intractable problem of the confetti. Passively powered by wave motion, the Beach allowed water to run over its surface, I was told, capturing the plastic confetti without the need for impractical filtering, and without catching marine life as well.

As the Golden Gate Bridge sank into the ocean behind us, Mary explained her position. She said it simply wasn't enough to talk about stemming the flow of plastic from land. Even if we stopped the influx from the United States, there would still be plastic from the rest of the world getting into the oceans. And she had spent her entire life on and around the ocean, building a successful sail-charter business. The ocean was her life's work. She felt she had to do something.

"So we have to work very vigorously to stop the flow," she said. "But we also have to effect cleanup."

Was that so wrong?

14 AUGUST—37°49′ N, 123°29′ W

We were elated to have set out, and relieved that the often-rough coastal oceans were forgiving that day. We watched the Farallon Islands go by—a set of remote, rocky outcroppings that, technically, are part of San Francisco. Then we were done with land. As if to announce it, a whale rose out of the depths, not fifteen feet to port. Staring down on its curling spine as it cut the surface and disappeared, we screamed with the exultation of inland people going to sea.

As with all true adventures, though, ours was to be remarkable for its long stretches of boredom. Soon our lives became an endless series of watches and off-watches—three hours on, six hours off, three on, six off, repeated ad infinitum—and I began to learn a bit about the seafaring life.

Our first duty on watch, unsurprisingly, was to watch: to keep an eye to port and to starboard for anything that might threaten to destroy us, other than boredom itself. During nighttime watches, I would stare into the darkness and try to see anything at all. On the second evening, tiny birds danced at the edge of our running lights, and I killed entire hours wondering if they were real.

The next task was the hourly boat check. The Pirate King would instruct one of us to walk the length of the deck, fore and aft, starboard and port; then to scout the belowdecks, to peek into the thundering oven that was the engine room (we remained under engine power even when augmented by sails); and finally to report back to him anything untoward or alarming, with special emphasis on whether the boat was sinking or on fire.

"Thank you," would come the Pirate King's approval, and we would turn to the main activities of the watch: the telling of stories and the sharing of bad jokes.

I was assigned to Watch B, which I rechristened Bravo Watch. It was, of course, the best of the three watches that made up the cycle. In addition to an official chronicler (me, self-appointed), we had Kelsey, a recent graduate from UC Berkeley, where she studied marine conservation; and both shipboard hipsters, Gabe and Henry, who were revealed to be old friends, inseparable since toddlerhood; and finally our watch captain, the Pirate King himself.

The Pirate King had turned out to be not only a hard-core example of seafaring masculinity but also something of a camp counselor. He seized every opportunity to teach us sea shanties, to recite poems both nautical and otherwise, to point out the constellations and unfold the mythology behind them, to show us how to splice rope and tie knots, how to braid special twine

"Turk's head" bracelets that would mark us forever as tall-ship sailors. On watch, as at meals, as in the lounge, he would break into story or song at the slightest provocation. I came to hesitate before taking a nap in the lower lounge, for fear that I would be awakened by the Pirate King, hanging upside down, splicing a lanyard with his teeth and singing a napping-shanty in twelve verses.

On our first midnight watch, he told us his life story: he grew up poor in Alabama, left home as a teenager, and remade himself in California. It was the tale of a young man enraged by how the world had treated him. Only through the trials and mortifications of sailing had he come to realize that, misfortunes aside, he could still make the choice to be happy. He had followed that revelation for the rest of his life, creating and controlling his environment, living for sailing. He worked as a captain and sailboat rigger and had lived on a boat since he was a teenager. The Pirate King had chosen his destiny.

Watch was also a time for gossip. Ships run on gossip, and it is the most reliable way to spread information among the crew. Boredom creates such a powerful suction in the mind for anything interesting, anything new, or anything related to your situation—the situation, that is, of being marooned on a small, steel island. Night watches, when the rest of the crew were sleeping, were especially productive. Entire shifts were spent reenacting the captain's social gaffes and speculating about whether Mary's goals for the voyage were achievable. We wondered how long the voyage would be, and mused about what, exactly, we were supposed to be doing.

The space between conversations, normally reserved at sea for quiet reverie and communion with the mysteries of the deep, was instead filled by Gabe, who for the duration of the voyage maintained a running series of food fantasies. Night and day, becalmed or in high seas, Gabe would welcome us into his inner restaurant, a sensual wonderland of Thai green curries and simmering stews and *more* green curries—always with the green curry—and hot liquored drinks to ward off the cold air that chased us almost all the way to the Gyre.

At times it seemed Gabe had no other way to approach the world. Once, during a discussion of the myth of the Garbage Patch as a "plastic island," I caught him staring into space, licking his lips.

"It's more like . . . like a thin minestrone," he said.

Oh, and then you also have to steer the boat, taking turns at the *Kaisei*'s tall, spoked wheel. You can pull off such feats of steering as you've never imagined: driving without being able to see in front of you (thanks to the masts, and the structure of the upper lounge, and whatnot), driving in the dark without headlights, driving in the dark without headlights while looking backward, with your hands off the wheel, drinking coffee, and telling bad jokes. These maneuvers and more, I personally executed.

All this is made possible by the absence, on the high seas, of anything else *but* the high seas. There is nothing to steer around, nothing to crash into, indeed no *things* whatsoever, except for you and your ship. If, within a ten-mile radius, so much as a rain squall or a tall wave threatened to violate our monopoly on thingness, the radar would sound an alarm.

All that mattered when you were steering, then, was the heading, which would be provided by the watch captain, in our case the Pirate King.

"One-eight-five," he would say.

"One-eight-five, *aye*," we would respond, duty-bound to get over the silliness of saying "aye" all the time.

You would then peer at the points of the compass as they wavered in the gimbaled steel housing of the binnacle, just beyond the wheel, and ponder how to make a five-degree course correction to a heading that wandered a good ten degrees back and forth, according to the swell and the wind and the whimsy of Poseidon.

16 AUGUST—36°55' N, 129°27' W

In the afternoon, we saw our first piece of debris. The honor went to Charlie Watch. Art, a retired science teacher both crusty and jovial, said he thought it was a net, although it was too far away to be sure.

Around eight thirty the following morning, I spotted Debris Item No. 2: a large bundle of synthetic yellow rope, to starboard.

After only three days at sea, the mind was already so tuned to the feature-lessness of the ocean surface that a bundle of rope was cause for major excitement. Even a fragment of kelp would have been a thrill, and this was actual *rope*. We went thronging to the rail. I wanted to cry "WHERE AWAY?" but restrained myself, as it would have made no sense. It had been me that spotted the rope, so someone *else* should have been crying "Where away?" so I could then call, "TWO POINTS ON THE STARBOARD BEAM!"

It was Patrick O'Brian syndrome. I had read too many of his rousing tales of early-nineteenth-century naval adventure. Now, on a large sailing vessel for the first time, I was afflicted by the urge, barely stifled, to scream "WHERE AWAY?!" whenever I had the chance, in rude imitation of the indefatigable Captain Aubrey. (The strange counterpoint to this urge was that I never got used to shouting "LAYING ALOFT!" before climbing into the rigging, as instructed by the Pirate King.)

The bundle of rope slid out of sight. "Where away?" I whispered.

There was more trash that day, small pieces here and there. We had no illusions that we were anywhere near the Gyre, though. We hadn't traveled far enough, and the weather was still cool and windy, not the warm dol-drums we could expect once we reached the Gyre. But it whetted our appe-tite. It sharpened our eyes. People began scanning the ocean surface for debris whenever they were on deck. Several people went up into the cross-trees to look out from above. The call came down of another rope sighting. (*Where away?*) Gabe and I went thronging to the rail. You must always *throng* to the rail, I felt, even if there are only two of you.

There it was: a tattered section of rope, maybe eighteen inches long.

"Oh, shit," said Gabe. "That is going to fuck up some ecosystem."

The sightings soon died down and by the following day the water was trash-free. We readied ourselves for its return. A logbook for debris was established—it lived in the wheelhouse, on the desk underneath the GPS/radar display—and a new task was added to our watch duties: debris lookout.

Two members of the active watch would sit in the bow, one looking to port, one to starboard, using a walkie-talkie to report anything they saw to a third member of the watch, who would note its latitude and longitude and the time of sighting in the logbook. The fourth watchmate would man the helm, and several times during the three-hour shift, at the word of the Pirate King, we would rotate.

I was underwhelmed by this debris-watch system. Mary had already said that this voyage wouldn't have the scientific focus of the previous year's, but if we were doing the work of debris watch anyway, it seemed a waste not to do it in a methodical or standardized way.

But no. There would be no real data collection, no pulling of nets through the water to quantify debris density at different coordinates. There wasn't even any consistent method for eyeballing it. Should we be looking everywhere and anywhere? Or should we be looking at a defined area, so that the debris count from one watch might be meaningfully compared with that of the next?

And how should different objects count? We would of course radio in large items to the wheelhouse. (*I've got half of a green plastic bucket. I've got a two-foot square of yellow tarp.*) But what about a two-inch fragment? A half-inch one? Only through the gradual buildup of a debris-lookout culture, transmitted orally from one watch to the next, did even vaguely standardized practices emerge. Our observations, it seemed, would be of little use to anyone else once we were done.

Soon, a pair of work lights were strapped to the netting underneath the bowsprit, and debris watch extended around the clock. Now we stood at the rails even at night, staring into the pools of light that trickled forward onto the rising, falling, onrushing ocean. In active seas, the prow of the ship became a mesmerizing twilight zone, where I stood watching bow waves crash aside and looked up at the *Kaisei*'s great square sails, taut against the night.

But when we couldn't find this reverie, some of us grumbled. What, exactly, was the point?

The point was that our goal was not to measure debris or to record it in any useful way, but simply to *find* it. We were looking for what Mary referred to as "current lines" of trash, narrow bands of high density. Mary spoke again and again of the current lines, and I saw that if we could bring the *Kaisei* back to port heavy with trash, it would validate the dream of cleanup. But for that, we would have to find the mother lode.

\\\\\\\\\

The impossibility of steering in a straight line is just one expression of the truth that at sea there *are* no straight lines. Nothing is level, nothing constant. Least of all gravity. You were once so naive as to assume that gravity was a force of uniform strength and direction. Welcome, then, to the *Kaisei*, where gravity is contingent, erratic, ever-changing. Just try putting down a mug of tea. All the flat surfaces of the world, formerly so useful, are now mere runways for your beverage, which will leap unbidden into the air and onto the floor, scuttling away in search of lower ground. For a mug—a book, a computer—to be left on a table or a shelf, it must first be restrained, like a lunatic strapped to a bed.

All the structures of daily life now find themselves built on quicksand. I go to the bathroom: I pee. The urine describes a sideways arc away from me (strictly speaking, I arc away from it), first left, then right, then left. Compound upon this the natural downward curve of the stream, and a newfound inability to stand upright, and now several integrals of calculus are necessary to ensure that a majority of one's piss doesn't end up on the floor.

The bathroom anointed, you make for the lower lounge. The world rotates. You plow against the left wall of the corridor, the right, the left . . . Soon you will realize that in fact you are walking a straight line, and it is the corridor itself that is driving the lane here, sometimes quite violently. It's up to you to shoulder off its aggressions.

Finally, you make it to the lower lounge for a quiet sit on the padded benches built against the hull. The lower lounge, like most places belowdecks, throbs with the vibration of the engines and the motion of the ship, with the parting of the ocean water as the *Kaisei* toils ever forward. Here, merely sitting, watching a movie on someone's laptop—*There Will Be Blood* stirring memories of Spindletop—you feel, more purely than anywhere else, precisely because you are trying to sit still, how the Earth's lines of force, once so parallel, so uniform, now swing and warp, bending the room into a haphazard, freaky place. One moment you are pressed against the cushions of the sofa. Then the pendulum of the world swings and you float half an inch into the air. An hour later, your face has melted and your stomach, having received so much from you over the years, now wants to give something back.

Will it never end? For three weeks, the very welds in the hull yearning upward and sideways?

You need your bunk. You stagger to the end of the glowering, thuggish corridor and back into your cabin, mumble something to the sleeping forms of your cabinmates, and then you are home, hidden away in the wooden womb of your bed, surrounded by clothes and blankets and bags of almonds.

I curled up, oriented so I wouldn't roll and crash against the walls as I napped, free for a moment from the need always to be bracing and balancing and holding on.

But even in my bunk I could feel the tireless ocean gravity, changeable as the wind. Under its sway, my organs and skin, my face, my mouth, they pulled against my skeleton: left, then right, then left . . . The thought surfaced that this was not, in fact, a special marine case. The boat and the ocean had not cast some churning and unnatural spell. They had merely revealed how the world really was. Gravity and orientation *weren't* reliable, except in the narrow instance of life on land. The worlds that sprang from the laws of nature were wavering and irregular. And so were our bodies—provisional, inconstant, flesh on a frame. And our lives and plans, too, oscillations in a medium, ripples passing up the swell.

18 AUGUST—35°46' N, 135°28' W

That afternoon, while I was at the helm, Mary came and stood on the bridge for a while. We hadn't spoken much since the beginning of the voyage. It was another odd effect of life on the *Kaisei*: thanks to the rigorous rotating schedule, you could see surprisingly little of someone who wasn't on your watch. Aside from meals—and even those were sometimes worth skipping for sleep—I might see the members of Alpha Watch only if I happened to be on deck taking pictures during their shift, or if there was a call for all hands to make sail. But Mary wasn't assigned to a watch and tended to be in her cabin when she wasn't visiting the on-duty watch or observing some debris being brought on board. So a certain distance built up.

Maybe I just felt awkward around her.

I shifted the wheel a few spokes to port, keeping course. Mary took a deep breath of ocean air.

"Been doing so much reading," she said. "Trying to synthesize every-thing and come up with the right approach." She told me she had a tall stack of books about ocean debris in her cabin.

How late, I thought. How late to be looking for the right approach.

She sat down on the edge of the bridge, leaning against the railing.

"So what do you think, Andrew?"

"Of what?" I said.

"Of life out here."

I considered the question. The sailing life is supposed to be the apo-theosis of freedom and adventure, but it seemed notable to me mainly for its indignities, and for the endless tasks, both awkward and arcane, on which our safety depended. It was like owning a house, but more likely to get you killed. The idea that sailing was an expression of freedom, I suspected, was merely a tool for self-soothing on the part of all the sailors and yachtsmen of the world. They had to justify why they bothered.

Mary was waiting for my answer, her eyes bright. I laughed. "Well, it's certainly different, Mary," I said.

She smiled and handed me a piece of chocolate. I thought of something she had told me back in California, advice for someone who had never been to sea. The trick, she'd said, is not to think of yourself as limited by the confines of your boat. You have to believe that you are limited only by the edges of what you can see *from* the boat. And the indignities of being at sea had let me realize the truth of this. The solution to every misery was to open your mind toward the horizon. To know that you were not on the ocean, but of it.

19 AUGUST—35°05' N, 138°42' W

On our fifth day—sixth? twelfth?—we got our first real taste. The air warmed, the clouds disappeared, the ocean became settled and smooth—and we caught the propeller on a ghost net.

Ghost nets are fishing nets and tackle that have been cast off or lost by commercial fishermen. As nets and their attached gear wander and float, they find each other, tangling into ungainly masses of net and rope riddled with fishing floats and other debris. They have the largest bodies of any species of nonlife in the Garbage Patch. Synthetic men-of-war, they continue to fish, entangling and killing animals as they roam the ocean for years, perhaps decades. And they're hell on propellers.

I awoke in my bunk to the sound of—what was that sound? Robin, another retired science teacher and a friend of Art's, was crouched next to me, nudging my shoulder.

You might want to get on deck, he said. The propeller got fouled on a net.

I realized what the sound was. After five days of constant motoring, the propeller was no longer spinning. "Where away?" I gurgled, tumbling out of bed.

We came on deck to find the Pirate King fresh from the ocean, stripped to the waist, droplets of saltwater glinting in his beard. I want to say he had a knife in his mouth. He had gone over the side to free the propeller. His quarry lay at his feet: a young ghost net, long and narrow, uncomplicated by other nets and ropes, not yet tangled into itself beyond recognition. The excited crew clustered around. We had only just reached the Gyre, and already the Garbage Patch had reached out, striking us a glancing blow!

I had my video camera with me and began recording Robin and his collaborators as they untangled the net. Mary was there, watching it unfold, oddly separate from the crew, as she always seemed to be. She picked up a corner of the net and turned it over in her hand.

"It's so hard to believe people throw stuff like this in the ocean," she said. I wasn't sure if she was talking to me or to my camera.

There was debris in the water again. Someone brought out a long pole with a basket of netting at its end, and we started hunting. There was a trick to it. If you left the net in the water, it became difficult to maneuver; instead, you had to stab the water just aft of an object as it passed. In this way, like Vikings spearfishing from the deck of a warship, we brought several scraps of trash aboard.

Meanwhile, Kaniela and Nick took the dinghy out, buzzing around to pick up bits of debris spotted by people aloft. Nick was on board as a representative of the Ocean Conservancy and was the closest thing we had to a professional marine biologist or ocean debris specialist. Every year, the Ocean Conservancy leads a gigantic volunteer effort called the International Coastal Cleanup, and this year Nick's efforts on the *Kaisei* would be the cleanup's symbolic beginning.

Bravo Watch began. I grabbed a walkie-talkie and took bow watch, calling sightings in to Gabe, in the wheelhouse, who would note them in the log. Nick and Kaniela also had a radio in the dinghy. If any of us saw something particularly interesting—a bucket, a large piece of tarp—Kaniela would gun the motor and the dinghy would skip across the ocean in hot pursuit.

I climbed out onto the bowsprit, watching the water stretch past, eyes peeled for plastic crates and buckets. The dinghy zipped forward with Nick in the bow, a figurehead in sunglasses.

Something was bugging me. I keyed my radio.

"Gamma whiskey breaker, this is bow watch alpha bravo comeback, over."

Bravo Watch liked its radios, and its nonsense.

"Loud and clear, bow watch," came Gabe's crackling reply. "This is the bridge. Can I get a two-five on your niner, over."

"Roger, bridge," I said. "Bridge, we are cleaning up the Pacific Ocean . . . *by hand*. Over."

Robin came forward to the bow to say hello. People liked to say hello when you were in the bow, not only because it was scenic and quiet but also because it was one of the few places where you could talk without being overheard.

I told him I didn't think our work was very useful.

"It's a joke!" he said, making a face. "The one thing is testing the ocean-current models. That's the one thing that could be real."

But as far as I could tell, the only thing Mary knew of the ocean-current models was a pair of GPS waypoints—one from NOAA and one from the University of Hawaii. Was that going to be it? Get to a waypoint and take a quick peek around? I had heard Mary say that her NOAA contact had suggested we "call him when we're out there." But now even that would be difficult. There was a satellite link on board, reserved for non-personal use—but it had stopped working before we even made it out the Golden Gate.

And what of "further testing collection technologies"? So far, we were innocent of any such initiative, except for Robin's project on the wheelhouse roof. He was working—at Mary's suggestion, I think—on jury-rigging a wide, rigid net that, were we to drag it through a dense swath of garbage, might snag a share. This was technology development on the *Kaisei*: a warm-hearted, wisecracking retiree gamely slinging a screw gun.

I had noticed something else on the wheelhouse roof. It was the Beach, stored from last year. This was the innovative wave-action device purpose-designed by Project Kaisei to isolate plastic confetti from the ocean water.

It was a slope-topped plywood box. Someone on the ship had built it during the previous summer's voyage. Now it was tied down just aft of where we stood at the wheel, steering the ship. For a long time I hadn't noticed it. Because it looked like a plywood box.

The dinghy zipped by on an intercept course for another scrap of debris. Robin reached out with two fingers together, as if he were going to pinch the ocean.

"It's like you're standing on the beach and picking up one tiny, tiny bit of sand," he said.

20 AUGUST—34°42' N, 140°19' W

In the middle of the night, I dream that I am at the wheel of a great ship, sailing the Pacific Ocean. The cold air is thick with moisture. The rigging creaks with the roll of the ship. Water hisses along the lee rail.

In the afternoon, Mary told us that we had passed the NOAA waypoint. It had gone by without fanfare, earlier that day or the previous night. Aside from the debris watch, no measurements were taken that I knew of. We had found no mother lode there, and so had moved on, setting course for the University of Hawaii waypoint.

The water, choppy and gray, was free of trash. We appeared to be having trouble finding not only a good stretch of garbage but also the North Pacific Subtropical Gyre itself. We were sailing on strong winds, which suggested that the high-pressure zone that is associated with the Gyre was farther west than usual. Would we never get to sail the seas of plastic? Mary maintained that there *was* trash in the water here, but that we couldn't see it because it had been pushed below the surface of the water by the increased wind and higher waves.

Nikolai Maximenko, the University of Hawaii oceanographer with whom Mary was working, later confirmed for me that this effect exists. But it was also increasingly apparent that day that the Garbage Patch was anything but uniform. It varied from spot to spot, heterogeneous and changing.

But that didn't matter. What mattered was getting farther west while we still could, and finding more trash.

21 AUGUST—34°44' N, 142°44' W

One week at sea. I spent half my waking hours dreaming of land. Land, which wouldn't move. A nice sidewalk, with the Doctor walking down it. The other half I spent in witness to the plain wonders of the open ocean and the numberless, fractal layers of its moving surface. Or I spent it out on a yard, wrestling a sail in the shining void.

We now lived for what we had all feared: going aloft. We waited for the call, edging toward the ratlines, ready to scramble upward, to the lower topsail, to the upper topsail, to the topgallant, perhaps a hundred feet above the deck. We edged out along the spars like arthritic monkeys, clinging to the yard with our bellies, as the Pirate King had shown us. I now depended on what I had dreaded would be the worst part of going aloft: the roll and pitch of the ship. I waited for the moments when it seemed to lift me upward in my climb, for the seconds when it glued me to the yard, when I could with confidence use both hands to tie a knot, unconcerned that I might be flung backward into blue nothingness, with only a short stretch of line to connect my waist to the rest of the world.

And I liked my shipmates. We were in that long moment of becoming friends, when the foreign and the familiar become, for a time, the same thing. Robin was revealed as a serial trickster, helpless before the opportunity to tell an off-color joke or to make joyous mockery of himself, of us, of everything. He told us old stories from his job, of being called into the principal's office—as a *teacher*—for some mischief inflicted on his students. Art, Robin's good friend, was a weather-beaten man in his seventies, with the thick brogue of a New England fisherman—yet he was a surfer and science teacher living in Hawaii. He climbed the ratlines and slid down the shrouds like someone much younger, hopping onto the deck, an ancient mariner flashing the *hang loose* sign. Then there was Adam, the shambling, culinary animal who inhabited the pitching metal box of the galley, fending off sliding pans and trays, deploying his ninja kit of cooking knives with focused abandon.

On Saturday evening we had a party. The concepts of "Saturday" and "evening" had long since lost their meaning in the constant cycling of the watches, but "party" was an idea that still held. It marked our entrance into the heart of the Gyre. At last, the air was warm, subtropical, the ocean glassier, almost smooth, the ship's deck glowing in the late sun. Joe, the engineer, cut the engine, and the *Kaisei* bobbed on the lazy swell.

Art climbed onto the roof of the upper lounge and improvised a modified passage from *Moby-Dick*, in which he evoked a search for the "great white ball of trash." We had seen a whale earlier that day, cause for some minor thronging. The truth in Art's joke, though, was that we would have been far more excited by a whale-size clump of trash.

Then Robin led us in a cheer, screaming something in Japanese, to which we responded with screams of *Banzai!* again and again, filling the ocean air with our cries. Only later did he admit that he had been screaming something, I think, about wanting to sleep with your sister.

Handles of vodka and rum appeared from hiding places belowdecks—it was supposed to be a dry ship—and the grog, a steel pot of fruit punch, was spiked and spiked again. Mary raised a mug, her face uncertain at where all this was headed, and made a few announcements. She reminded us why we were out here. It was about the plastic, about proofing the models, about finding the current lines. The data we were gathering was important, she emphasized.

Nikki, one of the more forcefully dedicated volunteers, chimed in. We needed to get more people on debris watch, she said. In her opinion, two crew members on lookout in the bow was not enough. She thought we should have someone aloft as well. "We need to find a way to *maximize our data*," she said, smacking her hands together.

Little empiricist flags shook themselves out all over my brain. Maximize?

Our observations were already of dubious scientific value. For one thing, the haul might vary widely based on how each watch went about its lookout work. Even something as simple as whether they faced forward from the bow, or out to the side, or aft, might suggest modulations in garbage density that didn't exist. And there were diverse opinions about how small an object could be and still count as an object, as opposed to a *bit*, or particle. While objects and fragments would be described in the log, and the time and coordinates of their sighting recorded, bits would simply be added to a running tally for the period of the watch. It had taken a good ten watches of debriswatching before consensus on such issues had coalesced. This achieved, our log of debris sightings, though quantitatively suspect, had a chance of some qualitative value, describing the ebb and flow of debris concentrations as we passed through them.

Now it was proposed that we *maximize* our data with additional lookouts. But this would throw the whole enterprise out of whack, if indeed it had ever been *in* whack. We would sight more of what was passing by the boat, and the log would show an increase, but the change would have nothing to do with a change in the water—only with how many people were on deck. Already, the log was showing the wounds of previous maximizations, in which Nikki had chosen to provide an extra set of eyes to someone else's debris watch. From the crosstrees, she had rained zeal and possibly duplicate sightings.

Kelsey, who had done her Berkeley thesis on marine debris, piped up before I did, pointing out that it was consistency that mattered, not a higher number per se. Nikki made an impassioned counterargument, centered on what a rare opportunity it was to be here in the Gyre. Then Art and Henry joined in, and Kaniela, and in this way, aboard the brigantine *Kaisei*, near latitude 34°36' North and longitude 143°21' West, at approximately 1930 hours, the scientific method was reinvented from the waterline up. Had there only been a high school science class present, it would have been one of the purest, most spontaneous moments of experiential education ever to unfold.

Empiric consistency won the day. The two-member debris watch was reratified, and the scientific community resumed its celebrations. By dark, we were sitting on the storage lockers on what Kaniela called the "poop deck," where we debated the etymology of the term, and whether this actually was one, and whether a non-poop deck could be converted into a poop deck by way of pooping, and so on. Then we were checking the sternlines Kaniela had set in hope of catching fish, and there were clouds of aromatic smoke, and we greeted every unfamiliar footfall with the paranoia of teenagers, even though some of us had not been teenagers for more than forty years.

I went to bed before it got very far. I had learned my lesson in Chernobyl, and was preemptively horrified at the idea of being hung over for our next watch, which started in four hours. So I missed the moment when someone realized that the fishing line had gone unnaturally taut, missed the moment when the monster was heaved on board: a mahimahi, easily three feet long, glistening and prehistoric. I slept through it all, slept through the commotion of Kaniela and George, the young assistant engineer, wrestling the incompletely killed mahimahi down the corridor, past my cabin, to the freezer; slept through the sounds of them mopping down the corridor, which even in their drunken state they realized had become a crime scene spattered with fish blood. I woke only for the watch change, every one of team Bravo late on deck, and one or two of us still badly drunk. At the helm, Kelsey responded to an order for a five-degree course correction with wild spins of the wheel to starboard, then compensated with even wilder spins to port. Walking forward to check the boat, I found George passed out with his fly open, lying on the netting below the bowsprit, his safety harness duly clipped in, a strangely beatific sprawl, the dream-like ocean flying by underneath.

23 AUGUST—33°36' N. 146°36' W

We were in it.

Nick raced back and forth in the dinghy, in full Ocean Conservancy mode, fishing out buckets and detergent bottles and jugs, laundry baskets and the odd hard hat. He filled out reports for the International Coastal Cleanup, typed the debris log into his laptop, and climbed up to the maintop to watch for large objects and current lines. It was infectious. You might feel suddenly alert and purposeful: Nick had entered the room.

The day's best catch was a large ghost net. George and the captain and I hauled it up the side, an ungodly tangle of net and rope and mesh, maybe three feet in diameter, that must have weighed at least 150 pounds. As it plopped on the deck, dozens of tiny crabs spilled from its recesses, flakes of cobalt blue that scuttled along the planking. They were the color of the Pacific. We threw them back. For all I know, such crabs only survive in the central Pacific if they have the animal of a ghost net to live in; and we, the destroyers, had pulled their host out of the water and consigned the survivors to certain doom in the crushing depths. We paid it no mind. In the future, ghost nets may be protected, just as whales and manatees are. But for now, it's open season.

I crouched by the ghost to inspect it. What did it look like? A brain? A jellyfish? A great mound of intestines? Ropes of every color and weave and composition and thickness knotted and twisted into one another. Several bright plastic lozenges—floats, or markers—lurked in the jumble, marked with Chinese or Japanese characters. Some of the knots in the ghost's component nets had clearly been tied by human hands, but others had to be the work of the ocean, tangled flights of topological insanity that bound one piece of junked rope or netting to the next.

Mary was watching. "I do hope we'll be able to show you something better than that," she said. She seemed to have little patience for the ghost nets, which were plastic-poor. It was plastic that she wanted, the current lines above all.

The current lines had become the Great White Ball of Trash prophesied by Art. Mary was confident that these strings of concentrated garbage, thrown together by the inner workings of the Gyre, were out here. The previous year's voyage had encountered them, she told us, and she was sure we could find one now.

I was getting tired of hearing about them. We were in the Garbage Patch—shouldn't we just be interested in what it was *like*? Instead, there was the sense that Project Kaisei only wanted the *stuff*. We needed something to show for our efforts. I wondered if this was symptomatic of a nonprofit bent on impressing its public or its funders. Would they be disappointed if we returned without a towering pile of trophy refuse? So we wanted more plastic, more dramatic densities, concentrations that we could really sink our teeth into. It was a more sophisticated way of believing in the plastic island, that idea that drove us all batty with annoyance. And I felt it kept us from appreciating the Garbage Patch as it was: just as vast and as problematic as we had expected, but deeply unspectacular. It required more than your eyes to grasp it. You had to think.

In this, the Great Pacific Garbage Patch is a cautionary tale in environmental aesthetics. We seem to require imagery to go with our environmental problems. If we don't have an image to be horrified by, we can't approach the problem in our minds. But sometimes the imagery distorts our thinking, or becomes a substitute for approaching the problem in the first place. And when there simply is no adequate image, we substitute others, creating islands where none exist.

No island, no carpet of plastic—yet we had without question entered the

Garbage Patch. We had sailed a thousand empty miles into nowhere, finally reaching this place. And what did we find here, so removed from humanity? Far more trash than you see in San Francisco Bay. More than you see in your own back alley. Every minute on the water, every thirty seconds, a bottle, a bucket, a piece of tarp, a sprinkle of confetti, multiplied by the countless square mileage of the Gyre. And yet if you looked across the surface of the ocean, it was unremarkable. Would-be debunkers need not resort to pointing out, as they do, that you can't find an image of the Garbage Patch on Google Earth. They should point out that you can't find images of the Garbage Patch *anywhere*.

This is because it isn't a visual problem, and this conflict between the reality of the problem and its nonvisual nature is at the root of the plastic island misconception. A metaphor is needed, a compelling image to suggest the scale and mass of the problem.

So let us explode the plastic island once and for all, and call it a *galaxy*. The Garbage Patch is like the Milky Way, an impossibly massive spiral that, because of its very vastness, is also phenomenally diffuse. Most of our galaxy is empty space. You could pass right through it without ever bumping into a star or a planet. The most massive object in the universe visible to the naked eye is made mostly of nothing.

If you were trying to figure out what a galaxy was, you would be plenty interested in the empty space between the stars, in whether or not it was truly empty, and in how the distribution of stars changed as you passed through the spiral arms. Like this, you might start to get an idea of your galaxy's shape, structure, and size. (Note: Your galaxy is many times the size of Texas.)

Similarly, if we had been dragging sample nets and taking real data, a stretch of empty Gyre water would have been just as interesting to us as one decorated with plastic, not least because access to the Garbage Patch is so difficult. In all of history, how many research missions had been to the Eastern Garbage Patch to study marine plastic? The folks at Algalita tell me it's about a dozen. The pool of existing data is therefore so small, and the character and dynamics of the Garbage Patch so poorly understood, that it felt negligent merely to obsess about finding the highest concentrations. But that is what we were doing. And if we were here to test cleanup methods, well, shouldn't those methods apply even in more diffuse areas? We were missing an opportunity to help inch the science forward.

And the science needed inching. A few hours on bow watch were enough to leave any thoughtful deckhand bursting with questions. Where were the plastic bags, for instance? On land, the Garbage Patch was often linked with plastic shopping bags. But here we saw no plastic bags. Were they below the surface? Had they broken up into small fragments? Were they all in the Western Garbage Patch, toward Japan? Or was it simply a myth that plastic bags make it out to the Gyre?

What about the stuff we *were* seeing? Where was it from? For most objects, it seemed impossible to tell. But there were more items with Chinese or Japanese words on them than with English, and a few with Russian, too. Anecdotally, this reinforced the idea that the Eastern Garbage Patch might be composed disproportionately of refuse from the *western* rim of the Pacific. Perhaps the Western Garbage Patch, the evil twin of the one we had now entered, was home base to material from the coast of the United States and Canada. If only we could have sailed another three thousand miles, I might have found all those Capri Sun pouches I went through in sixth grade.

It was also difficult to make even casual judgments about whether the trash we were seeing had come from land or from sea-borne sources. The common wisdom is that three-quarters of ocean plastic comes from con- sumer sources on land. This is borne out in places like Hawaii's Kamilo Beach, which catches the southwest edge of the Garbage Patch, and where lighters and toothbrushes and combs dominate. But for much of what we saw on the *Kaisei*, provenance was hard to determine.

And what factors determined what we could see? How, for instance, did an object's density and shape affect whether it stayed on the surface and how it traveled through the Gyre? And how old were the objects we saw? And how toxic? And what proportion was large objects versus confetti? And was there a class of sub-confetti particles, an as-yet-unknown kingdom of microscopic polyethylene flora? And most important, what kind of change did this wreak on the ecosystem?

Little of this is yet known to science—and to my nerd mind, it was the chance to help answer even one of those questions that should have been our white whale. It was a whale that swam alongside us for the entire voyage, but that we never noticed. And so the *Kaisei* sailed the ocean blue, irony on the wind, a mission to raise awareness, but not knowledge.

\\\\\\\\\\\

The Pirate King was a licensed ham-radio operator. Of course he was. He could have built a ham radio from an old soda can and a box of matches, underwater, while strangling MacGyver with his feet. In his circumnavigation of the globe, he had built up a network of land-based radio contacts, colleagues whom he had never met. Through them, he said, he could get a radio transmission patched into the phone system. We could call home.

I wanted to tell the Doctor I was still alive, and when we would return to land. But nobody knew. Although we were supposed to arrive in time for San Diego's tall-ship festival, Mary had been making indistinct noises about staying out as long as it took to find areas of higher trash-density. (Art's jokes about Captain Ahab were seeming less joke-like by the day.) The Pirate King, for his part, was bent on heading back. I couldn't tell if he was impatient with Mary, or tired of what he thought was a wild goose chase, or if perhaps he had a deep personal need to attend the tall-ship festival.

In the wheelhouse, the Pirate King keyed the radio and read the Doctor's phone number to an impossibly distant ham operator—a hobbyist in Florida, I think. Then he handed me the radio. I waited, while on the other side of the planet, a phone rang.

I never reached her. Several times I left a message, telling her the *Kaisei*'s latitude and longitude, and that I was alive and well, and that I loved her. She later told me the messages were sometimes garbled and unintelligible, my voice warped and splintered by its passage through the atmosphere. In those moments, she couldn't understand where I was, or anything I said. Only that it was me.

\\\\\\\\\\\

In the pit of night, the radar alarm sounded. A contact directly in front of us. The Pirate King said it was probably a squall, from how its profile on the radar screen changed and grew. Squalls patrolled this part of the ocean, hunched pillars of storm that could interrupt the night with lashing winds and rain.

In the wheelhouse, with our faces lit by the glow of the radar, we watched the contorted bolus of pixels bear down on us. It passed through the three-mile radius, then the two-mile. Then, slowly, it convulsed, stretched, and crept to port, passing within a mile.

We went outside and stared off the rail into the darkness, straining to see it. Nothing. No sky, no horizon. All night, we had seen nothing but a pair of stars, hesitant in the gloom. The radar said there was something out there, but we couldn't see it.

Then . . . something changed in our vision. Its outline came into focus. We *could* see it, faint and vast in the darkness, a monstrous anvil sliding over the ocean.

The sails hovered in the still air, indifferent. We went to bed.

24 AUGUST—32°59'N, 145°50'W

After ten days at sea, we turned back.

The tension among Mary and the Pirate King and the *Kaisei*'s captain had been growing for some time. All anyone knew for sure was that Mary wanted to stay out as long as possible, and that the Pirate King thought we needed to turn back, and that the captain liked to stage brief fits of nonsensical rage.

The Pirate King stood in the upper lounge and lectured us. He was as hell-bent on San Diego as Mary was on her current lines. As for the crew, we just wanted to know *when* we would head back, so that we could plan how we might, one day, return to our lives. But it seemed increasingly likely that we might wander the seas forever, a ghost ship in search of plastic. I saw Mary in the lower lounge, studying a distribution map in a textbook called *Marine Debris* (Coe and Rogers, 1996). "There should still be trash *there*," she said, pointing to a spot off Mexico.

In a pair of heated meetings, the argument finally spilled out into the open. The Pirate King insisted that we had to turn around right away. Not only was the tall-ship festival approaching—about which none of us really gave a damn—but Joe, the ship's engineer, was sick. He had some kind of throat infection, something that looked like it was getting serious. As an argument for speeding back to land, this was dubious; if Joe's condition was life-threatening, he would need an emergency airlift whether or not we turned the boat around.

But that was the argument that won the day. We *wore ship*, as sailors say, and headed east. Almost as soon as we had reached the Gyre, we were on our way out. Too many days wasted at the dock in Point Richmond, a little bad luck with the Gyre seeming to have pushed west that summer, and in the end I never got my turn in the dinghy, picking up Garbage Patch garbage with my own hands. And none of us ever went over the side to watch a ghost net swimming in its natural environment, attended by plastic minnows hovering in the spell of the fearsome, blue abyss.

\\\\\\\\\\

Bravo Watch was quiet that night. There were rumors that Mary was heartbroken to have turned back, that she considered it a major blow to Project Kaisei. It was impossible to know if such gossip was true. None of us volunteers were going to go knock on her door and ask. But it didn't matter. It was true in broad strokes. It felt like we had turned around as soon as we had gotten to the Garbage Patch. Had we even gotten to the heart of it? If we hadn't turned around, could we have found the current lines? Could we have found Art's Great White Ball of Trash?

It's sad how quickly a beginning turns into an end, with nothing in between. One day you still face an eternity at sea; the next day the voyage is over—though you may be days or weeks from land. It all depends on which way you're pointed.

We motored through the gloom. I was in the darkened wheelhouse, waiting to log any sightings from the generally fruitless nighttime debris watch.

Mary appeared next to me. We stood together, subdued, staring out at the night, at the murky silhouette of Kelsey at the helm, and listened to the engine drone.

I felt bad for her. The mission had been a great overreach. If our goal had been ecotourism—or pollution tourism—the voyage would have been a triumph. The Pirate King, aggressively self-righteous, never tired of pointing out the irony of us burning so much fuel to get out here. But that didn't bother me. People burn fuel all the time. They burn it to fly to London. They burn it to take a cruise. We had burned it to try to see something about the world. And though I was critical of Mary's goals, I could only credit her drive and determination. It was because of her that we were able to be out here, witnessing one of the great phenomena of our time.

I said some optimistic things. It didn't matter that we hadn't seen the current lines, I said. We had seen stretch upon stretch of particles. Places where they were too numerous to count. Places that prompted Henry to radio the bridge, "Oh, shit, they're everywhere." Weren't the particles the most intractable part of the problem, anyway? Hadn't we seen what we came for after all?

She murmured in agreement, unconvinced.

I watched the navigation unit. The radar echoes of nearby rain squalls crept across the display, primordial blobs of orange and yellow pixels that pulsed with a quiet, mysterious life.

"They look like little amoebas," I said.

Mary stared at the screen. A tear hovered at the edge of her eye.

"I wish they were islands of plastic," she said.

25 AUGUST—32°53'N, 143°08'W

The bowsprit was a good place for a morose crew member to cheer himself up. I sat on the netting, looking back at the place where the *Kaisei*'s prow sheared through the water. Looking down, I could see an area of water the size of a living room, undisturbed as yet by our onrushing hull. Hello, human-scale bit of Pacific. Goodbye.

The *Kaisei*'s mission had been easy fodder for a skeptic. It was the perfect expression of the weird symbiosis between an activist and the cause he or she is fighting against. It had been imperative for Project Kaisei to pinpoint, document, almost *celebrate*, the issue of marine plastic in its most horrifying instance.

But I wasn't so different. My mission was to find *the world's most polluted places*, as if I knew what that meant. Only if I found those ecosystems of despair would I be able to implement my conceit of contrarian ecotourism and compose my great elegy for the pre-human world. But instead of finding degraded ecosystems that I could treat *as though* they were beautiful, I was just finding beauty. The Earth had gotten there first. I went looking for a radioactive wasteland and found a radioactive garden. I went looking for the Pacific Garbage Patch and found the Pacific Ocean.

I sat on the bowsprit, leaning my face on one hand, a walkie-talkie slung around my neck, listening to the ocean crash against the ship. Soon, when we came closer to land, dolphins would find us, capering through the water below the bow net. We would lie in the netting, listening to them chatter and squeal. But for now, I was alone.

A plastic bottle ran under the boat.

I keyed the radio to report it to whoever was manning the debris log.

But before I could, a sprinkling of confetti appeared on the water, and then another bottle. Then some more confetti, a piece of tarp, some other objects—a crescendo of trash that peaked within a few seconds. I looked out to starboard and saw us bisect what I thought was a stripe of garbage several meters wide that ran toward the horizon.

It wasn't solid. No carpet of trash. But it was the densest, most localized stretch of debris I had seen all voyage. I called the wheelhouse on the radio and told them we had just crossed over a current line.

We didn't stop. Nobody even called *Where away?* Who was in the wheel-house—the Pirate King? The captain? They had eyes only for San Diego. But I had just seen it: the Great White Stripe of Trash. I keyed the radio again, filling with rage. This was fucking stupid, I told them. *I think we just crossed right over a current line.*

The *Kaisei* motored on toward San Diego. I think Mary was in her cabin.

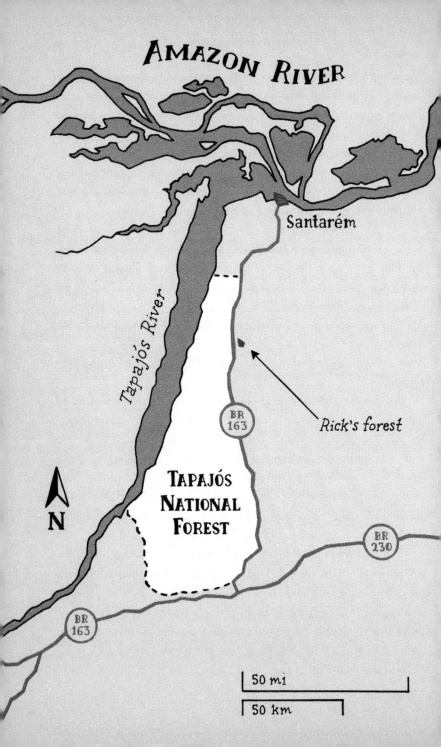

AMAZON RIVER

Santarém

Tapajós River

Rick's forest

BR
163

TAPAJÓS
NATIONAL
FOREST

BR
230

N

BR
163

50 mi

50 km

FIVE

SOYMAGEDDON

The smell of smoke, a plume rising by the road, and then we saw the flames. The forest was on fire. "Stop! Stop!" Gil cried, and Mango pulled over. Adam and I waded through the brush and emerged into a world of ash and cinder.

We were in the Amazon rainforest. The former Amazon rainforest, to be exact. The broad field where we stood was empty, freshly scorched to the ground. The air swirled with cinders. They mixed with sudden clouds of small, attacking insects. Where had they come from? Was there a species of bug driven to riot by the smell of smoke?

At the edge of the field, we found the fire crawling over what was left to burn. It reared up in brief flares, as tall as we were, then ducked its head back toward the ground. Adam started videotaping.

I turned back to the field, a monochrome square a hundred yards to a side. A single massive tree stood alone in the ruin. The ground was warm through my boots. In front of me, long bands of white crossed the ground, dividing, and dividing again, growing thin. They were the ghosts of trees. Felled and burning, they had turned to ash where they lay. Ashen branches sprouted from ashen trunks. I kicked one and it rose into the air, a white eddy circling on the hot breeze.

Everyone knows forests are good and deforestation is bad. Forests are habitat. Forests absorb carbon dioxide and forestall global warming. But not everyone knows that cutting them down and burning them not only releases carbon dioxide into the air but also creates local feedback loops that cause the forest to die back even further, meaning more habitat loss and more CO_2 emissions. The Amazon, at ten times the size of Texas, give or take a couple of Texases, has so much forest that to cut it back is to set off what some have termed a *carbon bomb*, with global consequences.

I had come to Brazil to see the burning fuse on that tremendous carbon bomb. There was only one catch: this probably wasn't it. You could even argue that this blackened, boot-melting wasteland, with its phantom trees and prowling flames, was protecting the forest from something even worse. Here, near the joining of the Amazon and one of its greatest tributaries, the people standing in the way of the rainforest's destruction sometimes looked a lot like they were cutting it down, or setting it on fire.

\\\\\\\\\\

India was supposed to be next. India, with the Doctor, married, for our honeymoon. But the Doctor had met me on the wharf when the *Kaisei* made San Diego, and called it off. A chasm opened in the ground and swallowed the world. It wasn't me, she said. But still. No wedding. No marriage. My life evaporated in a single afternoon.

I took it hard. The global environment, formerly such a candy store of problems, now lost its appeal. Even climate change and mass extinction seemed pretty minor next to the growing monument of my heartache. Yet books must be written. Who would take up the gospel of pollution tourism if I let it drop? I ditched India for the meantime and made for the Amazon, a fugitive from my own despair.

My friend Adam came with me. Or maybe I should say that I went with him. Ostensibly he was coming so we could shoot a television news piece in Brazil. We had collaborated on that kind of work before. But I also suspected that, unlike me, Adam *wanted* to go to Brazil. He may have thought, too, that I could use a little support. Left to my own devices, I might spend the entire Amazon trip in a hotel room, under a mosquito net, watching whatever passes for cable TV down there.

Friends—they're always trying to encourage you, and to convince you that you're not incompetent and unlovable and doomed to failure. Why can't they just butt out? On the other hand, with Adam on board, I could renounce all the detailed background research that I was going to blow off anyway. What was I going to do—crack open the Amazon with a week's googling? Screw that.

Originally, the Brazil trip was going to be about beef. Cattle ranching has long been a major driver of deforestation in the Amazon. Surely there was some friendly rancher out there who would give us the inside scoop on how virgin rainforest gets turned into hamburgers. Just think of the steaks we would eat.

But then we found out about soy. That's where the action was, we read (Adam read). Soy farmers were leveling great stretches of forest so they could sell animal feed to Europe. We ditched the ranching idea and chose Santarém as our destination. The city is the site of a controversial export terminal built by the multinational company Cargill to bring soybeans out of the Amazon. Near Santarém, we would be able to see it all: unblemished jungle, jungle being cut back, soy fields, and the terminal itself, a cruel agribusiness dagger thrust directly into the pulsing, green heart of the world. At least, this was my fervent hope.

With research outsourced to Adam, it fell to me to direct field operations. I put together a reporting plan.

1) Buy airline ticket (IMPORTANT).
2) Fly to Brazil.

3)   Exit airplane.
4)   Exit airport.
5)   Find taxi.
6)   Ask taxi driver to take us to the Amazon, preferably the part on fire.

This was a plan I could handle, especially once Adam—seeing I had no intention of addressing Action Item No. 1—went ahead and bought our plane tickets himself. (I still haven't paid him back.)

Then, as if to punish me for it, he shows up in my office wearing a green visor and waving a sheaf of papers, and gets all *NEWSFLASH* on me. The Brazilian government had just announced record-low rates of deforestation for 2010. The lowest rates of deforestation *ever recorded*.

Those bastards. Here I was, about to drag my ass to Brazil to go adventuring through a jungle-clearing orgy of absolutely first-rate proportions, and up pops President Lula to tell me that deforestation is more or less *solved*. It was disgusting. Since *when?* Wasn't deforestation like death and taxes? I'd been hearing about the inexorable destruction of the rainforests since I was a child. Now that, too, would be taken away from me?

It didn't matter. We had our tickets. And so we turned to the traveler's customary scramble of last-minute chores: suddenly you have a critical need for magnetic bug-proof socks, and a polarized hat, and a million other little purchases to help you convince yourself that you actually want to go on the trip.

While Adam spent his last few evenings researching deforestation patterns and Brazilian environmental policy, I screwed around on the Internet, disconsolately hoping for something to spur my curiosity. Somehow, I stumbled across an advertisement for some real estate near Santarém:

For Sale: 1,907 acres (766 hectares) of prime forest land adjacent to the Tapajós National Forest in the Amazon region of Brazil.

Rainforest for sale? I called at once. Soon I was talking to a gravelly voice on the other end of the line. His name was Rick, and he lived in Michigan, where he ran a business importing high-quality Amazonian wood.

"I own two thousand acres of what I consider the finest rainforest in the world," he said. "I made a lot of money in the exotic lumber business. So I bought it because . . . well, 'cause I could do it, I guess. It would be a soy farm or a cattle ranch by now, otherwise."

Since then, though, the economy had crashed, and business was bad. His company had shed most of its employees, and he couldn't afford to keep his rainforest anymore. And even though he had made his fortune in wood, he wanted to find a buyer for his forest who wouldn't just cut it down. Much of the land around it had already been converted to soy fields.

"I planned on making so much money in the business that I'd give my piece of forest to the state, or a college, or a nonprofit," he said. But for the past few years it had been a struggle just to hang on to it.

I told him I was going to Santarém in a couple of days.

Man, he said, I wish I were going with you. I was thinking about going, but I guess it's too late now. Have fun. You're going to love it down there.

\\\\\\\\\\\

A breezy city of a quarter million people, Santarém occupies a broad corner of riverbank just where the Amazon meets the mighty tributary of the Tapajós. Rick was right—Santarém is *nice*. I was instantly glad that Adam and I had decided not to go nosing into lawless pockets of the countryside in search of illegal logging. We might have overcorrected with tourist-friendly Santarém, but that was okay. I needed to relax, needed a vacation from the cold weather that had been creeping down onto New York, and from everything else.

We strolled out from our hotel in search of Gil, our fixer and translator. As we walked, we could see the meeting place of two massive rivers: on the far side, the muddy water of the Amazon, its opposite bank more than two miles away, and on the near side the Tapajós, famously dark and clear. Half a mile out from the bank, the two came together, mixing in a sinuous ribbon that looped downstream for miles.

Although Gil had come highly recommended, I was a little apprehensive about meeting him. The last few days had seen an erratic series of e-mails culminating with an abrupt request for electronics. "Bring me a iPod touch 4G 32GB," he wrote. "My girlfriend is pregnant if you bring me two i name our kid after you." Our fixer had turned into an Internet scam from Moldova. Uncertain what to do, I had split the difference and purchased a single iPod. It was now in my backpack.

Gil's place was an old yellow house just a few steps up from the waterfront. A windsurfing board lay on the sidewalk out front, and the house's door-like windows had been flung open to make a shallow terrace facing the river. Earsplitting rock music blared from inside. On tiptoe, I peered through a side window, looking for the mild-mannered face we had seen in a photo on his website.

Instead, I saw a half-naked wildman in surfing trunks, hunched over an old computer. I knocked. He didn't hear. He was glaring at the screen with a ferocious intensity, baring his teeth as he pecked at the keyboard. I knocked again, louder this time, and he looked up with a start, a flourish of hair tumbling down his shoulders. He had seen us. I felt the urge to run, but he had already sprung to his feet and was charging at us, shouting over the music. He was offering us caipirinhas—*Do you know what that is? It's our national drink! Do you have them in New York? You do?!*—and welcoming us to Santarém, and telling us how excited he was that we had come.

How to describe Gil Serique? He was a son of the rainforest, born in a speck of a village on the jungle bank of the Tapajós, a village without electricity or running water, accessible only by boat. "Paradise!" he called it. Now he was a multilingual river guide, a translator and fixer for visiting journalists, and above all, the prototypical Amazon beach bum. He windsurfed every day he could, launching into the river directly across the street from his house.

And he hustled. He had printed booklets to promote his guiding business, he cultivated contacts with the cruise ship operators who passed through Santarém, he obsessively updated his blog and his Facebook status (3,103 friends at last count). His house was a nexus for anyone interested in the rainforest, or its destruction, or in surfing, or in drinking, or in talking. He embarked at once on a series of absurd and exaggerated stories—about his neurosurgeon pilot friend who was descended from American Confederates, about having been in a Michael Jackson music video, about how, to his shame, he had once almost become an exotic-species smuggler.

The iPod Touch, it turned out, was part of a scheme to supplement his guiding income. Accepting it as partial payment for his services, he told us that iPod Touches were rare in Santarém; he planned to sell it at a 100 percent markup. "I should have everyone pay me in iPod Touches," he said. And then he told us some more about windsurfing. Always windsurfing. He was a man feasting on life like a crazed animal. If something was funny, or pleasant, or nice, then laughter was not enough. Instead his eyes would bug out, his teeth would flash, and he would emit a bloodcurdling scream. *"AAAGGHH!!"* He was a man afflicted with joy.

There were beers in our hands and we were leaning against the terrace, watching the good citizens of Santarém stroll the waterfront. Gil was the luckiest guy in the world, he told us breathlessly. He didn't know how he could be happier. He loved Santarém, loved the forest, loved his girlfriend, loved guiding, loved us. At forty-six, he was about to become a father. He shared the crumbling four-room house with his pregnant "bride-to-be." She was seventeen—not much more than a third his age.

"I've got the best life here, really," he told us, and then seemed to become overwhelmed by what he had just said. His eyes widened. "I love this town! I really love it! *AAAGGHH!*"

The conversation turned to deforestation and soy. It was disorienting the way Gil, with a maniacal gleam in his eye, somehow made screaming and drinking and cogent conversation all work together. He talked about the Cargill terminal, about the patterns of deforestation in the region.

"The roads in the Amazon were all built in the seventies," he said, his accent a reedy mix of Brazilian, British, and surfer. "Before that all the human pressure was along the waterway. With the roads, it's gone into the land."

The Amazon is a frontier forever under the sway of a new rush. Just the past hundred years have seen rubber booms, timber booms, gold rushes. Now soy and bauxite were taking the lead. But exploitation came in many forms. Even something as simple as boats with onboard refrigeration, which allowed fishermen to stay out longer, meant huge pressure on the fish in the river.

Would we like another beer?

In Gil's view, though, the exploitation had a flip side. "I got a lot more interesting work once they started devastating the Amazon," he said. "Normally, I'd be guiding nature lovers, but as Amazon conservation and everything gets bigger, I do more and more work with people like you, who want to see nature's problems." He was operating on the bleeding edge of eco-tourism.

But cruise ships were his bread and butter. During the season, ships packed with hundreds of tourists made their way from the mouth of the river up to Manaus, stopping for the day in Santarém.

"Cruise ship industry, this is a beautiful industry," Gil said. "Cause they come here, they're all old, and they're *loaded*. But they're cool—they're really cool. *Anyone* who comes to the Amazon is cool."

I had noticed a few tour-boat tourists turning circles on the waterfront, binoculars in hand. Cruise boats can ruin a town; in Juneau, where my father lives, the entire downtown area has become overrun with cruise ship passengers and the insipid economies that spring up around them. But that hadn't happened here yet.

Gil put down his beer. "I love whorehouses!" he cried.

I couldn't quite remember how we had come to this part of the conversation, but here we were.

He became serious for a moment: "I mean, the way we treat women in this country is *terrible*. Really." Then he brightened. "But still. There is this one whorehouse in Manaus . . . "

Was I supposed to nod? *Oh, yeah . . . Whorehouses!*

Our host went on, oblivious to our discomfort. He said you could get a girl for twenty reals—about ten dollars. Here was another form of exploitation in the Amazon, one Gil didn't mind participating in.

A pair of young women crossed into the park, strolling arm in arm. Gil assured us that the girls who walked by his house were the most beautiful girls in the world. "You're going to love the girls here," he said. "They're amazing." Concerned that he would start making introductions, we hastened to let him know that we both had girlfriends. Somehow, even for me, this remained true. For reasons unclear, I still lived with the Doctor.

It was dark. The park had filled with young couples, and families, and bands of teenagers. Children tumbled through the playground. A soccer game scuffled on the basketball court. Gil had turned philosophical. He felt so lucky to be alive, he told us. His sense of gratitude was oddly specific, though. He was grateful, first, for his carbon fiber windsurfing board. This was an amazing piece of technology.

"My first board weighed a ton, but this one is only like ten or fifteen kilos," he said. "I'm so grateful for that. And I'm so grateful that I'm forty-six and can get lots of little blue pills that can give me an incredible erection."

Adam choked on his beer.

"Not Viagra, man," Gil continued. "Viagra will give you a big erection, but it will make it very wide, you know." He gestured. "Very wide. But these little blue pills, I get them right around the corner. They cost like *nothing*. You take one of these pills, you will have a *serious* erection. You haven't taken those pills?"

We confirmed that we had not.

"So those are the two things I'm so grateful for at this time, that I can have in my life," he said.

"Carbon fiber and little blue pills," I listed, by way of a recap.

"That's right," Gil said, holding two fingers up. His eyes widened with realization. "*There's not a third thing! AAAGGHH!*"

We took off. We had hardly rested since New York. "Don't go!" Gil cried. But then he relented and walked us to our hotel, two blocks away. He had decided to go around the corner for a blue pill.

"I'm going to have some serious sex tonight," he said. "Thank you for this wonderful day!"

\\\\\\\\\\\

Highway BR-163 begins in Cuiabá, at the southern end of the Brazilian state of Mato Grosso, and runs north for more than a thousand miles, plunging directly through the Amazon. Built in the early 1970s, it is still mostly unpaved where it passes through the jungle, and during the rainy season it becomes a river of mud. Trucks founder in its legendary ruts and potholes, their progress slowed to less than a hundred miles a day. BR-163, it would be fair to say, is one of the world's crappiest major roads.

It has the distinction, however, of being one of only two roads that traverse the Amazon from north to south. As the *Economist* put it, BR-163 joins "the 'world's breadbasket' to the 'world's lungs.'" It links Mato Grosso—the agricultural powerhouse that has made Brazil the world's second-largest soy producer, after the United States—to the forested expanses of Pará. As such, the highway is a focus not only of commerce but also of some serious environmental anxiety. As Gil had pointed out to me, roads bring deforestation. You only cut down forests you can reach, and only turn jungles into ranches and farms if you have a way to carry off the beef and soy.

Once there is a road—even a crappy one—civilization begins to course along it, pushing out into the bordering forest. Humans like to think of themselves as builders and conquerors, but their presence spreads more like a vine, sending out tendrils, building a network, growing into the gaps until it forms a smothering blanket. Satellite images show that by the time it reaches the Tapajós River, BR-163 is sprouting cleared land in dense, perpendicular gashes, each a dozen miles long, like the teeth of a giant rake.

The road terminates, finally, in Santarém, at the western end of the waterfront. And it is here, not a hundred yards from where BR-163 runs out of places to go, that Cargill Incorporated of Minnetonka, Minnesota, built its soy terminal.

The terminal is Santarém's most conspicuous structure, a metal barn that stretches several hundred feet, with a huge Cargill logo on one side. A grain conveyor more than a thousand feet long extends from the northern end of the building to a tanker dock in the river. On the day we cruised past it, on a riverboat rented from a friend of Gil's, we saw a bulk carrier receiving its cargo, beans pouring out of massive downspouts, and plumes of soy dust floating out of the ship's compartments. Docked at the next moorage upriver was a Holland America cruise ship, a little smaller than the soy ship, and blinding with its bright wall of cabins. Next in line, smaller still, was a timber transport waiting for the cruise ship to vacate the dock so it could resume loading containers of wood. At a single glance, we could see the comings and goings of three major Amazonian industries: soy, tourists, and timber.

Construction of the Santarém terminal began in 1999 and was finished in 2003, even though Cargill had failed to do the necessary environmental impact study—a fact that resulted in the terminal being declared illegal by the Brazilian courts multiple times. Nevertheless, it opened.

For Cargill—the largest privately held corporation in the United States—building the terminal was a strategic move that allowed soybeans to get to market faster and more cheaply than before. Soy from Mato Grosso could be shipped north by river or trucked along BR-163. At the Santarém terminal, the soy could then be offloaded and stored before being shipped directly to Europe via the Amazon River. The fact of the terminal would therefore be a further incentive for the Brazilian government to pave the rest of BR-163. And that would mean yet another launching pad for assaults on the rainforest. (Paving a thousand-mile-long highway through the jungle, though, is easier said than done. As of 2012, it is still unfinished.)

There was incentive, too, for any farmer in Mato Grosso who cared to do the math. Why bother sending only your crop to the Cargill terminal when you could send the entire farm? Land was cheap around Santarém, and a soy farm built close to the Cargill terminal would save on both land and transportation costs.

Farmers flooded north. By 2004, a year after the terminal opened, cultivation of soy in the area had jumped to 35,000 hectares (about 85,000 acres, and a five-year increase of more than 2,000 percent) and land prices had shot up by a factor of thirty. Local farmers were under ever-increasing pressure to sell their land to the soy farmers. Soon, soy was being touted as rainforest enemy number one, and Greenpeace activists were crawling over the Cargill terminal building—in much the same way as they had crawled over the machinery of the oil sands mines—and demonstrators in the UK were showing up at McDonald's dressed as chickens to protest the use of Brazilian soy as chicken feed. The soy rush was on.

\\\\\\\\\

The highway in the dark. BR-163. The kilometers ticked by. This close to Santarém, the road was paved and free of potholes. Our driver sped south with abandon. He was a cheery, hulking man whose nickname translated as "Mango."

Locations on the road south of Santarém are found not by signs or named roads but by their kilometer number. We were headed for a turnoff somewhere in the low 70s. There, we would meet some people who spent their days ripping trees out of the rainforest. It was all perfectly legal, though—part of a sustainable logging project, and nothing to get upset about, unfortunately.

Dawn brightened by increments behind the tinted windows of Mango's car. We saw where we were. The rainforest. Right! After a couple of days in Santarém, a morose foreigner could almost forget that he had come here to see the jungle, to walk around inside the vast hydrological pump that is the Amazon, which lifts and distributes an ocean's worth of water across the Americas, shaping and driving weather patterns around the continent. Now, the Amazon canopy flew past, mist rising among the treetops. On the right, at the boundary of the Tapajós National Forest, trees approached to the edge of the road. Gil squinted through his window, looking for kilometer markers. On the left, the vista modulated between forest and rangeland—and then would fall away into the flat blankness of a soy field: mile-long rectangles of bare earth, stretching away to a residual wall of forest in the distance.

We reached the logging camp at around seven in the morning. The loggers were meeting in a bare, wooden room in the main building. Men and women in hard hats and work clothes stood in a circle and made announcements. There was laughter and applause. They joined hands and said a prayer. Then we went out and got into the back of a large, covered truck and bounced and shuddered down a rutted dirt road in the direction of the Tapajós River, into the heart of the national forest. We were riding with the Ambé project.

The idea of logging in a protected forest is probably abhorrent to most people, at least those who aren't loggers. After all, what's *protected* supposed to mean? Here in the Tapajós, though, a collective of people who live on the margin of the forest have been allowed a "sustainable" logging concession. The idea is that this will offer them alternatives to slash-and-burn agriculture and illegal logging, and provide economic development and improve living standards in the community without severely degrading the forest.

The key is that the people making money off the forest are the ones who call it home. Once they are living off it, they become critical stakeholders in its preservation; the community can only be sustained by the forest so long as the forest continues to exist. Suddenly it isn't just a few napping forest wardens who stand between the jungle and an army of illegal loggers and rule-bending soy farmers. The forces of *not in my backyard* are hitched to the cause of preservation.

The air changed as we entered the forest, becoming suddenly rich and earthy, the heat of the day eased by moisture and shade. The truck dropped us off and drove away, leaving us to follow a small survey crew on its morning rounds. I listened to the jungle: squeaks and whoops, squawks and trills, sounds that must have been coming from a bird or an insect but that sounded like someone blowing across the mouth of a bottle. Cascades of insect noise, almost electronic. Calls and responses. Sounds that were weirdly familiar—that I had heard before in movies and museum exhibits. The soundscape makes the jungle.

The survey crew went about its work. I tagged along behind a cheerful man with a machete, doing my best to stay out of the swinging whirlwind of his blade.

I wondered why I was sweating so much. In the deep shade of the canopy, it wasn't even hot—yet I sweated. Never in my life, not even in moments of optimal gym-nirvana, had I sweated like this. My shirt was soaked. My hair was soaked. My arms were soaked. I could not have been more drenched by a sudden downpour. The very hard hat on my sweating head was itself sweating. Moisture dripped from its brim. *How?* The question asked itself. *How does a plastic hard hat sweat?*

A needle of pain in my thigh. I looked down to see a green dagger sticking out of my leg. Its spiny brothers pointed at me from nearby branches. I pulled it out, a two-inch thorn. The air's suffusing odor of loamy decomposition suddenly took on new significance. It was the smell of the jungle breaking down and digesting anything that didn't keep moving. The Amazon wasn't just a lung. It was a stomach.

The morning survey done, we came back to the service road and walked along it for a while, toward a meeting point where the truck would pick us up. Every curve revealed a narrow vista—another towering queen of a tree, wearing a leafy corona over an impossibly slender trunk. A patch of brilliant indigo half the size of my palm materialized in the air: a butterfly. Adam crouched over a snail at the edge of the woods. In the middle of the road, a thin cable of succulent green hung out of the sky. I held it, felt the elastic connection between my hand and the distant canopy—and then gave it a tug. It broke, length after length of vine spooling down on my shoulders.

Gil was everywhere with the iPod Touch. Instead of selling it, he had fallen in love with the thing and had decided to keep it for himself. Now he roamed back and forth, taking videos.

Gil had a special connection to this place. His grandfather's family had lived here once, before it was a protected forest. They had made a settlement of their own, with about a dozen family members living off a piece of land that Gil's grandfather considered particularly rich. In the early 1970s, though, the government had decided to protect the area by creating the Tapajós National Forest, and had expelled many of the people who lived there. Gil's grandfather had been forced to sell his land.

"It was a reasonable amount of money," Gil told me. But it had been disastrous for the family. Instead of farming together, they found themselves looking for new and unfamiliar jobs. "Like truck driver, gold prospector, fisherman," Gil said. One uncle had opened a brothel and eventually sank into drug trafficking and violence.

Gil didn't think that creating the national forest had been wrong—only that it had been created on the wrong model. "See, in those years, the policy was based in the USA's Yellowstone," he told me.

He couldn't have chosen a more relevant example. Yellowstone was the first national park in the world, and its creation, in 1872, marked the moment in which white Americans truly fell in love with the splendor of the land they had conquered. But for that love to grow, the ideal of wilderness as a source of rapture and recreation had to be separated out from the loathing we all felt for native Americans, whose presence in the West tended to distract from our John Muir–style reveries.

Muir himself, the St. Francis of the American West and a prophet of wilderness preservation, admitted that he was barely tolerant of the native Americans he encountered. In 1869, he wrote that he would "prefer the society of squirrels and woodchucks." Muir's reverence for what he saw as the natural order of things continues to fuel conservation today, but it didn't extend so far as to include humans—of any color—as part of the environment. "Most Indians I have seen are not a whit more natural in their lives than we civilized whites," he wrote. "The worst thing about them is their uncleanliness. Nothing truly wild is unclean."

Native Americans were excluded from Yellowstone at its creation. Though people had been present in the area that was to become the park for thousands of years, native American practices of hunting and planned burning were anathema to a view of nature as sacrosanct from human involvement. If native Americans had been allowed to remain, they would have gotten in the way of all the nature white people wanted to appreciate. The creation of Yellowstone formalized the idea that human beings have no place in a protected wilderness—unless they are tourists.

As a result, some of the places we consider most pristine, most wild, are in some ways deeply artificial. A popular park like Yellowstone is probably more controlled, more managed, than the Exclusion Zone of Chernobyl. And even parks less besieged by visitors than Yellowstone or Yosemite are premised on ideas and laws that define human beings as outside of nature.

This artificial division between natural and unnatural pervades our understanding of the world. Industrialists may hope to dominate nature, and environmentalists to protect it—but both camps depend on the same dualism, on a conception of nature as something to which humanity has no fundamental link, and in which we have no inherent place. And it's a harmful dualism, even if it takes the form of veneration. It keeps us from embracing a robust, engaged environmentalism that is based on something more than gauzy, prelapsarian yearnings.

But we cling to the ideal of a separate and perfect nature as though to give it up would be the same as paving over the Garden of Eden. When I met with the writer and academic Paul Wapner, whose ideas I'm stealing here, he told me that a colleague had warned him not to publish his book on this subject, titled *Living Through the End of Nature*. His colleague thought it was a bad career move, and that anyone who argued that the concept of nature was no longer a useful one was giving away the farm.

The farm has already been given away. We're just so entranced by the concept of nature-as-purity that we won't face facts. Our environment is not on the brink of something. It is over the brink—over several brinks— and has been for some time. It was more than twenty years ago that Bill McKibben pointed out the simple fact that there is no longer any nook or cranny of the globe untouched by human effects. It's time to stop pretending otherwise, to stop pretending that we haven't already entered the Anthropocene, a new geological age marked by massive species loss (already achieved) and climate change (in progress).

But the dream of nature is so dear to us that to wake from it seems like a betrayal. The sense that we have not yet gone over that brink—not quite— is what motivates us to our ablutions, our donations, our recycling, our hope. But it is a great untruth. The task now, perhaps, is not to preserve the fantasy of a separate and pure nature, but to see how thoroughly we are part of the new nature that still lives. Only then can we preserve it, and us.

We went to find the rest of the loggers. The truck dropped us at the edge of a large, muddy clearing with a dozen large, felled trees stacked around its periphery. The air was alive with the riot of engines and saws. The clearing was a temporary holding area for trees that had been felled in the surrounding forest. A man with a chainsaw went from log to log, sawing off the sloping protrusions of roots at their bases, while other workers, both men and women, measured and marked them. An angry, saber-toothed fork-lift picked logs up in twos or threes and dropped them into a pile. They landed with a deep *thunk*.

After our peaceful stroll through the forest, the racket was overwhelming. To be honest, I think we were a little freaked out by how industrial it all was. I had expected a sustainable logging collective to involve a dozen nice folks and a good chainsaw. Instead, the nice folks had serious machinery and meant business. You could have taken pictures here that looked like every preservationist's nightmare—a mayhem of logs and mud. Or you could have taken pictures of the jolly, hardworking crew, and of the communities they supported, and of the forest that, it was hoped, their logging was helping to protect.

"The skidder is coming!" Gil said. "You can't see this very often! Let's go, let's go!"

We ran to the edge of the clearing and into the forest. A corridor of crushed vegetation led deeper into the jungle. Something had been through here. Trees were scraped and bruised where it had passed.

From the forest, we heard the shriek and growl of an engine. It heaved into sight: the skidder. This was how logs were brought out from the inaccessible interior, where they had been felled. They were dragged out behind this narrow, streamlined tank, a low, blunt-nosed hedgehog of a machine that was now headed our way.

Gil raised his iPod to record it. "We want to make sure not to be near it when it passes," he said, in the staring voice of the awestruck. The skidder plunged toward us, a colonizing robot from another world, surprisingly fast, shouldering trees aside as it bore closer, nearly on top of us.

And then we were running for our lives, screaming with joy and terror, leaping out of the way. It passed just a few yards from us, wheels grinding, and then it was gone. In its wake, a gigantic log slid coolly, massively, over the forest floor.

"Fucking *shit!*" Gil screamed. He was waving the iPod in the air. "It wasn't recording!" His disappointment took the form of an intense, quivering joy. Then we turned, and the machine was there again, back from the clearing, outbound for another log, bullheaded, inhuman, implacable.

On our way out, we stopped at the *patio*—the storage area near the highway, where logs awaited transport. They were piled twenty or more to a stack, each log three feet in diameter. We drove over soft ground flooded with rainwater, winding our way through a dozen stacks, two dozen. Flying ants wavered against the mountainous piles of logs. The purple stylus of a dragonfly appeared and disappeared. The air was thick with wood and rot.

Gil shook his head. "It's hard to believe this won't fuck up the forest, isn't it?"

\\\\\\\\\\\

Gil met us for breakfast at the hotel. As we planned our day over coffee and pastries, a muscular, middle-aged American man approached our table and started talking windsurfing with Gil. The American was wearing flip-flops and shorts, and had long, curly gray-blond hair and a deep, gravelly voice. A surfing buddy of Gil's, I thought. The subculture of Amazonian beach bums—one that I hadn't known existed two days earlier—was growing every day.

Then he turned to me, a business card in his hand. It was Rick. The man from Michigan who owned his own rainforest. On two days' notice, he had decided to come down to meet us in Brazil. He said there were a lot of misconceptions out there about the Amazon and about logging, and evidently he thought my presence in Santarém was a once-in-a-lifetime opportunity to get his story out.

I don't know what I had expected Rick to look like—a doughy guy in a polo shirt and khakis?—but it wasn't this. With his stony features and huge arms, he looked like a muscle-bound Gary Sinise. Or like someone who might choose to beat the crap out of the real Gary Sinise. He was accompanied by one of his few remaining local employees, a smart, understated young man whom Rick called Tang. They got some coffee and breakfast and joined us at our table.

Rick lived wood. His company imported wood to the United States, processed it, and sold it as exotic flooring. The business had been driven by the cheap money of the housing bubble, he said. "People building ten-thousand-square-foot houses because they could, putting in exotic hardwood floors because they could."

He got down to the business of misconception-correcting. "On TV in

America, they used to show some burnt, dying wasteland, and they'd have a logging truck driving through it," he said. "The assumption is that loggers cleared it. That they just nuke the place. But that's not the case."

Of all the trees growing in the rainforest, Rick told us, only five or six species were commercially viable. So logging in the Amazon had always been extremely selective. "If there were no cattle ranching, and no soy, the average person wouldn't be able to tell that one single log had been cut around here. Because there's no market for 94 percent of the forest."

Rick knew, though, that it was more complicated than that. "The worst thing loggers do is make roads," he admitted. And that created access for commercial agriculture. We later spoke to one of Rick's colleagues on this point. "Loggers don't destroy the forest, but they open the door," he said. "We are like high-class gangsters. We come into a museum, but we only steal the one multimillion-dollar painting. Then we leave the door open, and everyone else comes in after us, and they take everything. Even the lightbulbs."

Rick's problem wasn't with the fact of logging, but with how it was done. He couldn't abide waste. Huge amounts of wood had been wasted to achieve economies of scale. "It was so cheap here for perfect logs. It was the same in Michigan a hundred years ago. You'd lose money if you touched anything but the filet," he said, referring to the large, straight section at the bottom of the tree. The rest of the tree, from the lowest branch on up, was left to rot. "Billions and billions of board feet get wasted. I could build entire industries off the waste here, if I could just get access. It drives me crazy. I've been trying for years to see if I could get ahold of the tops left over by loggers—just their leftovers. And you can't do it. It just rots. There are so many rules, it's . . . " He grabbed his head. "It's Brazil." Sometimes entire forests were wasted. He had once visited a large bauxite mine nearby. Bauxite, the ore from which aluminum is derived, is big business now in the Amazon, and multinational companies cut down large tracts of forest to begin their open-pit mines.

"They had these piles of logs," Rick said. "They were prepping to *bury* them. It reminded me of pictures from Auschwitz. And can you get those logs? No."

He was so passionate about waste that he had started a Brazilian subsidiary based on it. The concept was to take leftover sawmill logs and use them to custom-build timber-frame houses, turning scrap into a luxury product. Rick nodded his head toward Tang, who had grown up nearby. "He's been building boats since he was three years old. He's one of the best timber framers in the world," he said. "So the idea was to use all that local talent that's here, and then use some resources that are being wasted. Not just turn the forest into a commodity."

He had called his local company Zero Impact Brazil. The lumberman was trying to turn over a new leaf. He admitted, though, that he had made most of his money on the commodity side: "For a while there, I was the largest buyer of forest products in Santarém."

Now that was all over. The housing boom had crashed, and the market for exotic flooring had gone with it. The entire timber industry had died back. Tang told us that over the last five years, two-thirds of the sawmills in the area had closed. Logging trucks had gotten scarce.

I stared into my coffee cup. Let this be a lesson to you, I thought. Never wait to see a rainforest being logged out of existence, because one day you'll wake up and it will be too late.

"Yeah," Rick said. "Lots of money got dumped in here from all over the world. Big investments. They come here with real big eyes." And like so many others, they had gotten burned. "The typical business model in the Amazon," he said, "is you go there with a lot of money—and you leave broke."

Now it was Rick's turn. The timber frames weren't selling. Zero Impact Brazil was surviving only by selling off its assets.

We stood up to go, agreeing to talk again soon, to arrange a visit to Rick's rainforest. We'd go down there and "goof around," he said. He was insistent on that point—on the goofing around. Adam and I exchanged a glance. What did that mean, exactly?

Rick also wanted to talk some more about the forest, about waste, about his company. "I wanna portray us as at least the guys who have got good intentions," he said.

\\\\\\\\\\\\

Stoking a mild despondency about Brazil's failure to keep up its end of the environmental-horror-story bargain, I turned for succor to the Catholic Church. Adam had uncovered an activist priest who promised to say inflammatory and pessimistic things about the Amazonian situation. He had made headlines overseas—the BBC called him "the Amazon's most ardent protector"—and had a reputation as a fierce champion of the rainforest.

Gil knew where to find him, of course. He knew everybody, perhaps because he spent his every spare moment on the tiny terrace of his house, greeting passersby, waving, hollering, gossiping. Walking around Santarém with him was like tagging along for a victory lap with a popular former mayor. Acquaintances and friends shouted from windows and sidewalks on every block.

We went looking for Father Edilberto Sena not at his church but at the offices of his radio station, which says something about his approach to liberation theology. The station operated from a small, two-story building on a busy street up the hill from the river, and Sena used it to promote his activist causes, beginning with an editorial broadcast every morning.

From half a block away, Gil spotted him pulling into a parking spot, and we introduced ourselves on the sidewalk. He was a short man, youthfully sixtysomething, with a pugnacious smile and good English.

As we walked toward the entrance of the radio station, two young women crossed our path. Sena stopped in his tracks and turned to us.

"One problem of the Amazon . . . " he said. "Too many beautiful girls around."

Smiling, he laid a hand on his chest.

"A poor priest *suffers*."

From a media relations point of view, this seemed like a questionable way for a priest to start in with a pair of visiting journalists. But it was part and parcel of Father Sena's rebel persona, which he clearly held very dear. In his office, I asked him what he thought about the Brazilian government's figures, which showed that deforestation had reached record lows.

"Bullshit!" he cried, his face shining. He acknowledged that deforestation had diminished in 2010, but insisted that this wasn't the whole story. "When you put it together with the deforestation of 2008, 2007 . . . " He chopped his hand against the desk. "For the last eight years, we have a sum of *16 percent* of the Amazonian forest destroyed."

I was feeling better already.

Unfortunately, his figures were badly exaggerated. It had taken more like thirty years, not just eight, to destroy 16 percent of the Amazon. But that was beside the point. Deforestation was only part of the story, he said. "We ask, 'Why are you, Mr. Government, continuing with huge projects of hydroelectrics in Amazonia?' Government has a plan to build *thirty-eight* hydroelectrics in Amazonia." There were even dams planned for the Tapajós. "I feel the contradiction from the government," he said. "Saying they are fighting to stop deforestation, and at the same time they are planning to build hydroelectrics that will destroy rivers, forests, and the people."

Sena had brought the same defiant spirit to the fight against soy farming in the area. The organization he founded, called the Amazon Defense Front, had partnered with Greenpeace to protest the Cargill terminal. But the collaboration didn't last.

"Greenpeace was a very important ally from 2004 to 2006," he said. "Then we stopped . . . Our styles were different. We went to the street, to make protest. Greenpeace went jumping on the roof of Cargill." He laughed. "And filming, and showing to Europe and to the world that *Greenpeace was here!*"

There were philosophical differences, too. "I am not an environmentalist!" he said, waving his finger in the air. "I am an *Amazonianist*. Because the Amazon is more than the environment. It is also the people."

He smiled the smile of a firebrand. "Greenpeace has money. But it doesn't help much when you don't have a holistic viewpoint. They defend the forest. They defend the animals. They forget that the environment includes the people that live here. That's the difference. We defend our people."

It was the Ambé approach, applied to environmental politics. Without taking people into account—in your activism, in your national parks— something essential was missing. And Sena didn't just mean indigenous people. He also included the small farmers who had been displaced by soy, and more than twenty million other people spread across the Amazon basin, whether in the countryside or in big cities like Manaus. They were all critical stakeholders.

But there was at least one group that didn't count.

"Before 2000, we didn't know the plant of soy," Sena told us. But by 2001, soy farmers from Mato Grosso had started showing up. "They went with money and bought this land," he said emphatically. "They didn't come to live here. They came to cultivate here."

Newly arrived from the south, the soy farmers had not integrated well, not least because their mechanized farms offered few jobs for the people of Pará. The locals took to calling the soy farmers *soyeros*, a play on the Portuguese word for dirty.

"Soyeros don't like it when we call them that," Father Sena said. "But they *are* dirty. They didn't come here to join us, but just to suck the possibilities of this land."

Call it the Sena Doctrine. People are an indispensable part of the environment—unless they're dirty bastards.

\\\\\\\\\\

We trundled down BR-163 in Mango's car again, to about kilometer 45, where we met Nestor, a small-time farmer who had survived the soy fever and kept his farm. Nestor sold us beers and, together with his son, took us on a walking tour of his manioc fields. "There were many people living here who owned small farms," he said. But in the first five years of the decade, buyers from the south had swarmed in, bidding up land prices. Most people had taken the money. "They sold the land, and the tractors came and finished with it all." A nearby village called Paca had been wiped completely off the map to make way for soybeans. Even the Pentecostal church in the village had sold out and moved. "They sold it all," said Nestor's son, laughing. "They brought down the church to plant soybeans. You can't even tell there was a church there."

Nestor blamed the local politicians who he said had brought Cargill in: "The government brought these people to bring progress. And maybe it did. But it also brought bad things . . . People saw the money and thought it would never end. One person would sell, and that would inspire the next person to do the same."

It sounded like a frenzy, I said. Gil translated, using the word *locura*, for "madness." Nestor and his son nodded vigorously. "Era," they said. *It was.* Along this stretch of highway, Nestor told us, only he and his brother had kept their plots of land intact. Everyone else had sold at least part.

The frenzy had changed the local environment, in ways both subtle and obvious. We met multiple farmers who complained about the chemicals that neighboring soy farms used on their crop, and about how the soy monoculture had increased the burden of pests on small farms nearby. "There are a lot of diseases in their fields," one man said of the soy farmers. "I plant rice and I get nothing. If I plant beans, the insects eat it all. We can't harvest anything." He claimed that the soy farmers were able to thrive only because of all the fertilizers they used.

He said, too, that such large, open tracts of land changed the winds and the temperature around them, and that the simple absence of shade made life harder. Where once they had walked great distances in a day's work, the wide expanses of the soy farms meant less protection from the punishing Brazilian sun—and thus less walking.

We asked Nestor why he hadn't sold. Buyers had been offering big money. He said that wasn't important. He didn't like money.

"If you don't like money," I said, "then we won't bother paying for the beers."

He laughed. "We like a *little* money."

Now the ones who had sold their land and moved to Santarém regretted it, he said. They wanted to come back. Another small farmer down the road told us the same. "Many think that when they move to town, the money they got will never run out," he said. "They go to town, buy a house, a TV set, a refrigerator. But they never got an education, so they can't get a job. When the money runs out and they have no means to work, they regret selling the land."

We never stopped hearing about the families who regretted selling—from Nestor, from other farmers, from Father Sena. Here, people worried less about soy's effects on the forest than about its effects on their society, about the ways it had impoverished the people who had sold their farms.

"Now they are after a small plot of land and can't find one," Nestor said. "Their daughters became prostitutes. Their sons became glue-sniffers."

Gil said it was the same as when his grandfather had been bought out of his home in the Tapajós National Forest. "They ran out of money right away. It happened to most of my uncles."

\\\\\\\\\

Another interesting thing about Nestor was that his farm was on fire.

Much of our conversation took place in the middle of a smoldering field, similar to the one in which I would later melt the soles of my boots, staring at those ghostly, tree-shaped piles of ash. The fire was the reason we had stopped to talk to Nestor in the first place. I was here to see some deforestation, dammit, and if a field of slashed-and-burning trees wasn't deforestation, then I didn't know what deforestation was.

Turns out I didn't. In the Amazon, deforestation is a dispiritingly messy subject to unpack. Even Adam found the topic surprisingly opaque once he got down to the nitty-gritty of it for me. The main theme of any in-depth article about deforestation in Brazil, he once told me, ought to be how frustrating it is just trying to figure out what counts.

Take Nestor's case. You would think a charred stump is a charred stump, but not so. Nestor was just rotating his crops. *Slash and burn* has a scary ring to it, but around here slashing and burning is often part of a farmer's yearly routine. The piece of land Nestor was burning had already been cultivated multiple times. He would grow a crop of manioc—a root vegetable known elsewhere as cassava or yuca—and then leave the field to become overgrown with trees and brush while it lay fallow.

Now, several years later, he was going to cultivate it again. To prepare it, he had cut down the new growth, let it dry for a few weeks, and was burning it off. From a carbon point of view, his footprint was neutral: the CO_2 going into the air on the day we visited was CO_2 that had been sucked out of the air by this vegetation over the course of the past five years or so. True, there was a carbon debt—and habitat loss—from the original establishment of his farm, but that had been decades ago.

The real argument is over what drives *new* deforestation. And it's not as simple as who's holding the chainsaw. A person cutting down trees might be there because of government incentives to encourage the settlement of "undeveloped" areas. A soy farmer may only have come north because land was too expensive in his home state—or because an American buyer like Cargill has set up shop in Pará. All sorts of things can prompt deforestation at a distance. A soy farmer using previously cultivated land could argue that he isn't destroying the Amazon. But what if the small farmer who sold him that land goes off to clear new land somewhere else? To whom do you attribute the destruction?

Even if you can answer that question, you are then confronted with the situation that once an area of rainforest is settled, the settlers themselves become the de facto caretakers of whatever is left. Landowners in Brazil are subject to a unique forest law that obligates them to leave 80 percent of their land in native forest. Even giant soy farms aren't allowed to clear more than 20 percent of their land. (The farming lobby is trying to change this law.) If the law were effective, it would mean that anyone who cut down twenty hectares of jungle would end up being responsible for protecting another eighty.

It's hard to imagine a muddier picture. Decades ago, when Nestor first set up his farm, he might accurately have been characterized as the face of deforestation—sucking the possibilities of the land, as Father Sena would say. But now Nestor was a local stakeholder whose livelihood as a farmer depended on resisting the waves of development that followed him. His permanence on the land had earned him a place under the Sena Doctrine. But wouldn't that happen to anyone who stayed long enough?

Come back in thirty years. Maybe there will be a proud *soyero* making a stand, refusing to sell his farm for the construction of a mega-mall next to the Tapajós National Forest, and we'll call him a defender of the Amazon.

\\\\\\\\\\\

"I don't know what you do around here after dark," Rick said. "I don't drink. I guess if you drink, if you like to party, you can go to a bar and visit with people."

Adam and I had bumped into him in the park across from our hotel and invited him to eat dinner with us. At an outdoor restaurant across from the waterfront, we sat on plastic patio furniture and ate steak and chicken, and Rick pressed on us once again the need to visit his forest. "We can swim, we can goof around," he said.

Rick had first come to Brazil twenty-five years earlier, seized by the idea of importing wood directly from Brazilian suppliers. In an era before e-mail or widespread fax machines, finding those suppliers had meant coming down in person. So that's what he did, wandering from city to city through the Amazon, knocking on sawmill doors, even though he spoke no Portuguese. (Twenty-five years later, he still didn't.)

It hadn't taken long for the sawmill operators to figure out that, although he "looked like a hippie," as he put it, Rick wasn't there to protest, or to chain himself to a tree. He wanted to *buy* trees.

It made him his fortune. He became a major exporter of wood from Santarém. He told us that for several years in the 1990s, he was the biggest customer of Cemex—at the time, the largest logging company in Santarém. The world's appetite for exotic lumber had been one of the forces sending tendrils of destruction into the rainforest, and Rick had cut out the middlemen, and fed it.

Yet he seemed less a businessman than a searcher of some kind. Whether it was the experience of seeing his business die back, or something else, he had been humbled.

He showed us a photograph of the river on his phone. Underneath the distant sliver of a kitesurfing kite, a tiny figure rode the surface of the water.

"That's me," he said.

He put his phone away. "You know how some people say that when you're surfing, you connect with the water, or whatever?" he asked. "I can kind of relate to that now. When you're kitesurfing, you're really in touch with the environment. You've got the water, and the waves, and also the wind. You finally relax, and stop trying to control it. You stop fearing it."

He laughed at himself. He was a gruff chisel of a man. Adam and I sat and listened. Over our shoulders, the Amazon and the Tapajós mixed and flowed, invisible in the dark.

"I don't know what you'd call that," Rick said. "Something like a religious experience."

\\\\\\\\\\

We went to find the *soyeros*, those dirty bastards from the south.

"I found out land was cheap in Pará," said Luiz. "It was the only place I could afford it. So we came here to buy a plot of land and own it. That's why we're here."

Luiz was a short man in his early sixties, with watery eyes and an uncertain gait. He was a soy farmer, with three hundred hectares under the plow, just up the highway from Nestor's land. He was also, to my eye, drunk.

"Would you have moved here if the Cargill port wasn't here?" Adam asked.

Luiz frowned and shook his head as Gil translated. "What would I do here?" He had come for the same reason as the other soy farmers. He had realized that while the price of soy would be the same in Pará as in Mato Grosso, the cost of transport would be much less.

"We only came here because of Cargill," he said. "Not that Cargill went to Mato Grosso and called us. But we watch the news."

We walked along the edge of his field, deep and crumbly with muddy earth, to the barn where he kept his combine. Luiz plunged his forearm into a sack of grain and pulled out a handful of dry soybeans, his balance wavering as he held it up for us to see. "Soybeans are dollars," he said.

Luiz could see me staring at the combine, a tall, old machine with green sides. He swung up the ladder to the driver's perch, and soon the machine rumbled to life, its rows of harvesting blades gnashing and turning. He turned it off and I climbed up to the steering wheel. I peered out at the soy field in front of me, and imagined rumbling through it on the combine at harvest time.

Things hadn't worked out perfectly for the *soyeros*. The value of Luiz's land had crashed by 60 percent since he'd bought it. Even worse, when he'd bought it, he hadn't known he wouldn't be able to clear off all the trees.

"The *environmentalists*." He spat the word. *Ambientalistas.* "They came with these laws, and it was forbidden to clear more than 20 percent of the area." He had been forced to lease additional land in order to grow a large-enough crop. It made no sense to him. This was rich, flat land. It ought to be cultivated. And the forest on his land wasn't even virgin forest, he said. There were no good hardwoods left on it, no monkeys, no fruit. The law ought to be that if you're going to protect a forest, it's a *real* forest.

But that wasn't how it worked. "For the environmentalists, the farmers of Pará are criminals, some sort of thug," he said, and laughed. "They'd be more hurt to see a smashed tree than a dead farmer."

It wasn't just the environmentalists either. Although religious, Luiz had stopped going to church. "I stopped going because I would feel angry," he said. He knew what people like Father Sena called him. He just didn't understand why. "The priests attack us, but we're not criminals. We're not harming anyone's lives."

We left. In the car, speeding back toward Santarém, Mango laughed. He couldn't believe Luiz hadn't known why people hated the *soyeros* around here.

I'll tell you why they hate you, Mango said. It's because you're cutting down the forest, you asshole!

\\\\\\\\\\

First there is a toucan on a branch, minding its business. The sound of synthesizers. Then a magnificent tree, rising skyward behind the toucan. There are shafts of sunlight. Look how they filter down, extra Amazon-y.

Then, grinding through the vegetation—*a bulldozer.* It crashes toward us, a mechanized demon in the Garden of Eden. Closer and closer it comes, filling the screen. The image fades, and now we see the results: a wasteland of burning stumps and blackened earth. The sky is orange with smoke and flame. It's the image Rick complained about: that loggers *just nuke the place.*

And then, among the devastation, a lonely figure. Head downcast, clothes scorched and torn, he wanders the destroyed forest. Who is this desolate stranger, this angel of grief lost in a nightmare of deforestation?

It's Michael Jackson. Obviously. And he is here to ask penetrating questions about the environment: *"What about sunrise?"* he sings. *"What about rain? What about all the things that you said we were to gain?"*

Sales of *Earth Song* were a little weak in the United States, but in Britain it was Michael's most successful song ever, hitting No. 1 for six weeks over the 1995–96 holiday season. And it has the best environmental music video of all time, meaning it's totally wretched.

Soon we see sad-faced African bushpeople staring woefully at a murdered elephant; and sad-faced Amazonians in traditional undress, watching helplessly as trees fall in the rainforest. There are also some sad-faced Croatians thrown in for good measure. It was the nineties, after all.

Michael falls to his knees, pounding the earth with his fists. Then the Amazonians, the Africans, and the Croatians fall to *their* knees and begin pawing at the ground. Soon, everyone is grinding their fingers through the soil, shaking fistfuls of dirt at the sky.

A mighty wind begins to blow. (They actually show the planet from space, engulfed by the mighty wind.) Michael is now in full Christ mode, standing with arms outstretched, holding on to two twisted tree trunks to keep from being blown away by the righteous hurricane he has summoned. And then—wait for it—*time begins to run backward*. In Africa, the elephant resprouts its tusks and hops up, newly unmurdered. Michael Jackson and the downtrodden peoples of the Earth are *undoing all the damage*. All hell unbreaks loose. "*Where did we go wrong?*" he screams. "*Someone tell me why!*" The meaningless lyrics are paired with images of meaningless fantasy. Smokestacks suck their own filth out of the sky. In the Amazon, two local lumberjacks look on in astonishment as their work is undone, a massive tree lifting magically into the air and rejoining its stump. We cut to a close-up of a logger's awestruck face and see—

It's Gil.

We paused the video. On the screen, video-Gil stared up at the magic un-logged tree. Next to the computer, real-life Gil stood with a gleeful I-told-you-so look on his face.

"*AAAGGHH!*" he screamed.

All week, he had been spinning his story about having been in a Michael Jackson video, but we had never considered the possibility that it was actually true. Finally I had called his bluff—and there he was, on YouTube, intercut with the King of Pop's righteous convulsions. The young Gil Serique, son of the Tapajós, with more than ten million views.

\\\\\\\\\\

Cargill said we could visit. It had taken a week of phone calls and e-mails to exotic places like São Paulo and Minnesota to convince them we were harmless. The only reason they relented, I think, was that we told them we were shooting a television piece for an American news program, which was true.

It felt like a get. We had secured access to the Amazonian terminal of the largest private company in the United States, the driver of the Santarém soy bubble. This was ground zero for the destruction of the Amazonian rainforest, a match held to the carbon bomb's fuse. In terms of habitat destruction and climate change, this was the temple of doom.

Or not. Adam, the ingrate, says that I can't say any of that stuff: it's *not true*. This is the problem with having colleagues with integrity. They're always bringing it to work.

Soy, he tells me, whether or not I want to hear it, has never been a dominant cause of deforestation in the Amazon, has never been responsible for more than a tenth of the destruction. A measly tenth! The soy frenzy in Pará had generated a lot of heat in the media and among environmentalists. But when you look at the Amazon as a whole, soy has never come close to matching the deforestation caused by cattle ranching. In fact, even slash-and-burn farmers like Nestor still account for more deforestation than soy ever did. Which means that maybe I should have cast Nestor as a villain (even though he was friendly and sold us cheap beer) and been more sympathetic to Luiz, even though he had stumbled around shouting like a drunken jerk.

Why, then, all the ruckus about soy? The answer, perhaps, is that soy burst onto the scene with such frightening speed—and also that, in Cargill, environmentalists had found a concrete target.

In 2006 Greenpeace released a report called *Eating Up the Amazon*, which gave Cargill a lot of attention. The report traced soy grown on deforested land through the Cargill terminal and all the way to Europe, where it ended up as animal feed for chicken and beef sold in McDonald's restaurants. This crystallized the problem in a powerful way. After all, an activist who can cry *"J'accuse!"* toward a specific McNugget is an activist who has nicely focused the case. Furthermore, the McNugget connection provided two strategic choke points for Greenpeace to attack: the Santarém terminal and the McDonald's boardroom.

To the terminal, they sent their ship the *Arctic Sunrise*, which blocked the dock and delivered a team of activists who climbed up onto the works, as they do, briefly shutting it down.

To McDonald's, they sent the heavy artillery: people in chicken suits. In Britain, Greenpeace shock troops dressed as poultry danced through McDonald's franchises and chained themselves to restaurant tables. The news footage of this is pure Dadaist entertainment, with a police officer approaching one of the chickens to ask who's in charge.

Activists should break out the chicken suits more often. Within weeks, the pressure had worked its way backward along the supply chain. Cargill came to the negotiating table, along with all the other major buyers of Brazilian soy (including companies such as ADM). The companies were clearly terrified that they were next in line for their own visit from the chicken suits.

Not three months after it all started, the soy buyers signed an agreement under which they would buy no soy from recently deforested land. According to David Cleary, a strategy director at the Nature Conservancy, which brokered the deal, the agreement goes beyond the standard set by the Brazilian government, which allows 20 percent deforestation on farmland. Under the terms of the agreement—known as the *soy moratorium*—Cargill won't buy soy from any farm where a single tree has been cut since the moratorium began.

In contrast to the Ambé project, which works from the bottom up, by way of local stakeholders, the soy moratorium is a top-down approach that depends on technology. The way it works is that soy growers must register their land, and Nature Conservancy and Cargill staff show up and walk the perimeter of each farm with handheld GPS devices. The farm is then monitored by satellite for any deforestation that occurs within its limits. Brazil already had a very sophisticated system for monitoring deforestation—it depends in part on information from NASA—but without knowing exactly which land belonged to which farmer, the government couldn't do much about it. Now, though, they can monitor each specific farm to make sure trees aren't being cut down, and cheap GPS technology makes it remarkably inexpensive to graft this additional monitoring onto the existing satellite-based system.

The crazy thing about the soy moratorium—aside from the role that people in chicken suits played in creating it—is that it actually seems to have worked. It is still in effect, and soy-driven deforestation in the Santarém area has stopped dead. I know this because Adam showed me a graph, based on Brazilian government data, that shows the region's cumulative deforestation. Immediately upon the implementation of the soy moratorium, the line goes flat.

Luiz, the soy farmer, testified to its effectiveness. "If you're not operating legally, you can't sell a single grain of soy," he groused. "You have to be legal, or Cargill won't pay you." If it weren't for the moratorium, he told us, "we would plant everywhere."

Luiz was pissed off about it, but from a conservation point of view, the agreement had been so effective that there were now hopes to apply a similar system to the problem of cattle ranching. If successful, it could prove to be a major innovation in controlling deforestation in the developing world.

I couldn't stand it. Was there no end to the good news? Let's recap a few of my most unwelcome findings:

a) Amazonian deforestation tour conducted precisely at time of record-low deforestation.
b) Topic of soy as a journalistic focus revealed to be a mistake provoked by a passing fad among the environmental media.

c) I therefore fail to address the real problem, which is still beef.

d) Multinational corporate villains revealed as key participants in an anti-deforestation success story.

e) Goddamn it.

And let's not forget that the only people I saw tearing down trees were possibly angels of sustainability and local empowerment; and that the people setting forests on fire were friendly small farmers. It was hard to know where the catalog of disappointment would end.

In any case, we had our access. Adam and Gil and I showed up at the Cargill terminal, the temple-of-not-so-much-doom, and were ushered into an air-conditioned reception room, where we awaited the attention of the terminal manager. In a display case at one end of the room, a glass goblet of soybeans stood next to a collection of bottled cooking oils and jars of mayonnaises and other food products derived from Cargill ingredients.

The terminal manager walked in, the commander in chief of the million-odd tons of soybeans that passed through the terminal every year. A solid busybody of a man with thinning hair and a green polo shirt, he already looked impatient.

The great thing about the tour he gave us was that even though we had a newfound appreciation of how Cargill might conceivably be helping to *mitigate* deforestation, and even though that meant that any story we did might actually take a *positive* spin on the company, this didn't stop us all from playing out our appointed roles. We were the journalists out to get the multinational company, for which the terminal manager was a bloviating mouthpiece.

After some blather about "environment is a priority" and "safety is a priority," he told us we would not, as promised, be able to go out to the dock to see grain pouring into a ship bound for Liverpool or Amsterdam. Nor would we be allowed inside the huge hangar of the grain storage area. That is to say, for reasons of safety and lameness, we wouldn't be allowed "in" the plant at all. Nor would we see so much as a single bean of soy, beyond those in the reception room display case. What was this, an oil sands bus tour?

Instead, the terminal manager led us on a circuit around the outside of the storage facility, pointing out the truck bay—here, too, safety was a priority—and other completely boring, soyless locations and features.

On a stretch of wet concrete between the water and the storage building, the terminal manager stopped and turned to us.

"Here we have a forest for the preservation of native trees," he said.

We looked around. What was he talking about? To our left there was a small triangle of grass with a dozen scrawny trees. Only two or three of them even truly qualified as trees. The rest were little more than half-naked sprigs sticking out of the ground. This was their forest? "We have had some difficulties in growing the trees," he said. "But we take very good care of them."

He stood there with his hands on his hips, and we stared. Here, in the heart of the Amazon, we had found the most pathetic nature preserve in the entire universe.

"It's small," the plant manager said. "But it's symbolic of our commitment to preserving the forest."

\\\\\\\\\

At kilometer 77 on BR-163, Rick's rental car turned onto a side road. Mango followed, guiding our car off the pavement and onto dirt. We turned east, leaving the Tapajós National Forest at our backs. A wooden sign with hand-painted letters stood on the highway shoulder:

VENDE-SE
766 HaDE
MATA VIRGEM
4km—>

"For sale . . . Damn, I wish I had some money!" Gil said, watching the sign go by. "Shit, I love *forests!*"

We were driving to Rick's rainforest. At long last, we were going to find out what he meant by *goofing around*. Adam and I each had our own worst-case scenario, and they both involved kidnapping.

Two kilometers in, and the fields and overgrowth on the right became a thick secondary forest—the dense growth that floods into a previously disturbed area. Because secondary forest lacks the shady canopy of untouched, primary forest, it often grows with a thicket-like abandon absent in primary forest.

Or as Gil put it, "Shit, look at this fucking forest! That's a *serious forest*."

Rick's land was taken care of by a young man named Antonio, who had been born and raised on an adjacent small farm. Antonio's family home was a long, single-story cabin made with rough-hewn planks of itauba wood. It sat on a low rise bordered by trees. Children scurried and played in the yard. There were chickens, and a well, and an outdoor kitchen where Antonio cut fruit for us with a machete. It seemed vaguely like paradise. Only if you had been here thirty years could you understand that this was a landscape in the throes of change.

It was Antonio's father, Raimundo, who had established the farm in the 1970s, as part of a wave of settlers encouraged by the Brazilian government, which was high on the idea of developing the supposedly unproductive land of the Amazon. New arrivals could get a 100-hectare (250-acre) plot for almost nothing.

"It was beautiful then," Raimundo said, sitting in the front yard of what I might have otherwise called his still rather beautiful homestead. "The forest was vast, full with everything," he said. "Game and all living beings. In those days everything was easier."

Like Nestor, Raimundo was now suffering the effects of the most recent wave of deforestation, and he echoed Nestor's complaints about the large soy farms. But Raimundo didn't hate the soy farmers. He had been offered a trunkload of cash, and had turned it down, and that was that. "We feel good, because everything they do, it's for Brazil," he said. "But what can I say? We feel the heat, because of the cleared land."

"Once it's cleared, it will never be the same again," Antonio said. "We know for a fact that place will never be what it was before."

And it wasn't over yet. We asked Raimundo what he thought the area would be like when his son reached his age.

"If they don't come up with a law for a man to protect the forest he lives in, there will be nothing left," he said. "Nothing left." Perhaps because it was so poorly enforced, he was unaware that such a law already existed.

Rick's cabin was back in the forest, perhaps ten minutes by foot from Antonio's house. The path led through the woods, along a wooden walkway that passed over a shady, clear-running creek, and finally to a sandy clearing. The cabin was a simple structure, no more than a few bare rooms made of planks cut by chainsaw. We slung our hammocks on the narrow porch.

A wasp was harrying Adam. As he tried to squirm and jump away, it became enraged and stung him on the cheek. "What did I do wrong?" he asked himself. Then, looking at the encroaching jungle all around, he drew the lesson. "The forest is my enemy," he said.

We dumped our bags in the cabin and gathered in a troop facing Rick, our commander. "Are you ready for your jungle adventure?" he growled.

Rick had himself never made it to the depths of his own forest—because of how large the property was, he said. But Tang suggested it was because Rick just kept going around and around on the same trail.

"Have you seen the lake?" Tang asked.

"No," Rick said.

"Have you seen the field?" Tang asked.

"No." Rick smiled ruefully. "I've probably only seen fifty hectares of the place."

The highlight of our walk through Rick's rainforest was a magnificent tauari tree. Its base spread out along the ground in huge triangular fins that embraced cavernous spaces perhaps twenty feet tall. It wasn't a tree so much as a group of searching, wooden walls that had come together to build a minaret.

Rick stared up at it. "As you can see here, this thing is like an art piece," he said. "Thousands of trees like this have been cut. Millions, probably. Tauari is a commercial species. Most of it went to France. For some reason they love it. Europeans love tauari."

With his fist, he pounded on one of the giant, fin-like roots. It made a deep, thudding reverberation.

It was a spectacular tree, mystifying in its beauty. And yet, standing under it there in the jungle, I saw that I would have to stop fighting a realization that had been dogging me the whole trip: a rainforest, however fascinating, is still just a forest.

This is not as vapid an observation as it sounds. The legend of the *jungle* is so powerful, and so laden with the importance of biodiversity and the lungs-of-the-planet thing, that we forget that an Amazonian rainforest has an awful lot in common with a regular North American forest. To wit: it is a forest. Yet the Amazon of our dreams persists—a place overgrown with mythology and legend, with humid stories of explorers and murky tales of pre-contact tribes. You almost expect it to be made of jade.

This is true even when the mythology is negative. Werner Herzog, in a wonderful interview during the making of his movie *Fitzcarraldo*, proclaimed that the jungle was full of "misery," that the birds cried out not in song but in pain, that the Amazon rainforest was a world of obscenity and horror. But in this, Herzog was being no less mawkish than Kathleen Turner in her breathy search for a giant emerald in *Romancing the Stone*—not to mention Michael Jackson in his *Earth Song*. Then there's James Cameron's *Avatar*, the ultimate expression of jungle-as-magical-place, driven by a story so painfully condescending to its forest-dwellers that he could get away with it only in science fiction.

In these cinematic Amazons, sunlight must always filter seductively, a leopard or a giant spider—or a fetching blue alien with breasts—must be around every bend, and every step on the path must be won with a machete slashed through the succulent fronds of something greener-than-green. Poison darts fly unceasingly from blowguns, leeches latch instantly onto legs and bellies. And piranhas, of course—always piranhas—wait for the dip of an unwise toe in the river. It's not just a jungle. It's Eden with some danger thrown in.

Maybe other parts of the Amazon are like that, but around here, it was primarily a forest. It had trees, and leaves, and dirt, and animals. And in this case, it had an owner. Most important, it also had a swimming hole. Finally, Adam and I understood what Rick had meant by *goofing around*. He had meant there was a rope swing.

The swimming hole was down the path beyond the cabin, where a stream—a tributary of a tributary of a tributary of the mighty Amazon— eddied into a wide pool surrounded by trees. A small wooden swimming dock had been built out into the water.

The Americans had brought their swimming trunks, the Brazilians their briefs. There were the requisite jokes about piranhas, and we dove in. Rick climbed the slanted trunk of a collapsed tree, holding the rope swing in one hand. He surveyed his kingdom—and then jumped, carving a magnificent arc, his mane of gray-blond curls trailing behind him, a late-career Tarzan in board shorts.

If anything, his arc was a little too magnificent. It brought him over the platform of the swimming dock, and for a moment I thought he was going to break his neck. Instead, the Michigan-born Lord of the Jungle watched with a bemused grimace as the dock passed underneath him, and then he swung back, still seizing tight to his vine, his feet dragging through the creek, and at last he let go, collapsing ingloriously into the water.

\\\\\\\\\\

We were on an island in an ocean of soy. Out at the property line, Rick's forest fell away into a huge, flat expanse of dry earth. We had gone to take a look.

It wasn't yet planting season. Heat wavered over the crumbled dirt. A trio of silos stood in the distance. Gil danced back and forth taking video with his iPod Touch. Rick pointed out the line of green running alongside the field. It was the border of his forest. He had owned it for ten years.

"This huge, thousand-acre soybean field here, at one time was all forest," he said. "One year I came through here with some people, and there was a huge pile of logs, still burning. They just cut that piece out."

It wasn't just the small farmers who had felt the pressure to sell, but anyone who owned uncut forest in the Santarém area. When I asked Steven Alexander—another American who owns a tract of uncut forest in the area—whether anybody had offered to buy his land, he laughed.

"I had a line of people trying every day to buy it!" he said.

A gentle, white-bearded man in his early seventies, Alexander had been living in the area for thirty years, working for a health and education NGO, and later as a forest guide. He had bought his land back when it was cheap. Now it, too, was an island, and he took a dim view of the Amazon's long-term prospects.

"My guess is that it will become much like North America or Europe," he said.

"Really?" I asked. "The Amazon will look like France?"

He thought about it for a moment. "Over a period of time, a hundred years, two hundred, I don't think we can expect to see anything more than preserves . . . Everything around that will go." The Amazon rainforest would remain as a mere archipelago, islands of protected forest scattered across the river basin.

"It seems to be the way of the world now, doesn't it?" Alexander said. He smiled gently. "More and more people, more roads, more development, less forest. That seems to be the trend."

\\\\\\\\\\

Cargill, Greenpeace, and the Nature Conservancy all agreed that the soy moratorium was a success. But it had left some business unfinished. For one thing, there was the question of the Cargill terminal's dubious legality. And what about the small farmers who, having sold out, found themselves profoundly impoverished? Both of these concerns had been fundamental to activists' case against the soy industry. Greenpeace had produced a short film—titled *In the Name of Progress: How Soya Is Destroying the Amazon Rainforest*—that highlighted those two issues in stark, accusatory terms. But once the moratorium was signed, they were dropped.

This, more than anything else, explains the rift between an international NGO like Greenpeace and an impudent local activist like Father Sena. In his view, Greenpeace and the Nature Conservancy had secured a weak agreement. The decrease in deforestation, he thought, was due to the global economic slowdown, not the moratorium. And even if the moratorium was stopping soy farmers from cutting down forest themselves, what about the small farmers they displaced? They were much harder to track. Meanwhile, nothing was being done to mitigate the damage that had already been done—and the Cargill terminal was still allowed to exist.

"Greenpeace forgot about us," Sena said. "They used our movement." They had made heroes of themselves, declared victory, and moved on.

When Adam later tracked down Andre Muggiati, a forest campaigner at Greenpeace Amazon, he nearly admitted as much. "Edilberto is still a good friend," he said of Sena. But he said that, for Sena, "the only solution for the problems would be to put Cargill out, to send all the soy farmers back to the south. That is not reasonable. We always knew that at some point we would have to sit at the table with Cargill to get an agreement. If you ask the impossible, you never get to a solution." Activism could only do so much. "Capitalism and free initiative are legal in Brazil," he said. "You can't come to Cargill and say, 'Go away.' You cannot go to the soy farmers and say, 'Return the land to the peasants.'"

Who could disagree with Muggiati? But however sensible, his words could have come out of Cargill's own mouth, and so hint at some uncomfortable parallels between the agribusiness giant and the environmental NGOs that opposed it. Both sides had operated with a degree of realpolitik about the possibility for justice in the wake of the soy boom. Both had maneuvered through legal gray areas to advance their cause. Of course, Greenpeace had no hand in overturning the way of life that had sustained the small farmers of Pará. But it did use them as poster children in its campaign, only to discard them once a realistic political goal had been reached. And the goal it achieved—if indeed it was truly achieved—was to protect the forest, not to address the social ills that had gone along with the soy boom. It's also difficult not to find some irony in a guy from Greenpeace invoking realism and the rule of law—when a good deal of that organization's public activism depends on idealism and on the targeted flouting of the law.

Father Sena—militant priest, spitfire idealist, girl-watcher—ended up on the outside. When negotiations for the soy moratorium began, Sena's Amazon Defense Front had been part of them. But the ADF wanted too much: a ten-year moratorium, extending two years retroactively, instead of the more realistic, yearly-renewable arrangement that was ultimately agreed upon. Sena told us that he had walked away. "We said, 'Forget it. You can cheat people from the United States, but you cannot cheat us.'" That left Pará's soy moratorium to be designed by Cargill (of Minnesota) and the Nature Conservancy (of Virginia) and Greenpeace (founded in Vancouver) and a grab bag of other NGOs and agribusiness giants, most of them from the northern hemisphere.

The soy moratorium may prove a great success story in the end. It may even herald a way forward for the control of deforestation. But it lacks exactly what the Ambé's sustainable logging project hopes to establish: local players who have a stake in the forest's preservation. In the case of Ambé, the very people benefiting from the forest's exploitation have a profound incentive to do it sustainably. But the soy moratorium's several constituencies are different. One—Cargill and its competitors—is at best indifferent to the rainforest. Another—the soy farmers—would cut it down if they could get away with it. Yet another—the mostly foreign-based NGOs—can only hope to build their moral imperatives into the machinery of agribusiness and development, through political maneuvering and legal cajoling.

Finally, there's whoever is still left in and around the forest—the people who, for whatever reason, didn't sell and haven't cleared all their land. And who's to say how long it will be worth their while to hold on to it?

\\\\\\\\\\\\

In the evening we ate dinner in Antonio's yard and then retired to Rick's cabin in the rainforest. From the clearing, I watched bats transit the moon as it rose over the jungle. Howler monkeys groaned in the distance. Nearby, a troop of frogs set up a ceaseless knocking rhythm, anchoring an aural tapestry of peeping and piping and cricketing, cicada-like sounds that glimmered in the darkness. Adam thrashed around with a flashlight in his mouth, dosing himself with choking clouds of bug spray.

At the edge of the clearing, Tang produced a guitar and began strumming, singing a plaintive tune into the dark.

"What is that, Tangy?" asked Rick. "That something from your home village?"

"It's Dire Straits," Tang said.

Gil had passed out in his hammock, a lumpy pod hanging between two trees in the dark. We walked out past Antonio's house, past a pair of drowsy cattle, to where the soy fields began. Tang lay on the road, his arms behind his head, and Rick and Adam and I stared at the night sky. Our original hope had been to see the distant glow of fires in the south. In the days of the free-for-all, Rick told us, it had been possible to see the night sky aflame with apocalyptic color, the radiant flush of a forest casting off its earthly bonds. The awestruck way he spoke reminded me of Hilton Kelley's description of refinery flares in Port Arthur. There is a kind of destruction that has beauty in its weapon.

Tonight, though, there were no horizons of orange and red. It wasn't really burning season, if they even bothered to have a good burning season these days. And so we were left with silent flickers of lightning in the far heat, and the stars. The last time I had seen the stars so well had been on the

Kaisei, listening to the Pirate King digress on Orion, on the Pleiades, on Cetus.

A large bat flapped out of the night and passed over our heads. "Here they come—agh!" Adam cried, ducking for cover. Even in the middle of a soy field, the forest was out to get him.

The bat followed its erratic flight path out over the soy field behind us. I looked at the field, how it stretched out of sight in the dark. Just how did they clear this stuff?

Years ago, Rick told us, a rancher down the highway had bought a large piece of forest and wanted to clear it. "He hired five hundred guys, bought five hundred chainsaws, and just went at the forest," he said. "In one season, I don't know how many thousand hectares he cleared, just *brrrrcchh!*"

Rick had said he wanted to be portrayed as one of *the guys that have got good intentions.* And I thought his enthusiasm for the rainforest was genuine. But the fact that he had been a major exporter of wood from Santarém also meant that his business almost certainly had been built on illegally logged wood. As recently as the mid-2000s, 60 to 80 percent of the wood coming out of the Amazon had been logged illegally. And although Rick didn't like to get specific, more than once he had made vague references to the frenzy of the old days, of the crazy things he'd seen in the Brazilian logging industry. He knew the business he had made his fortune in. When he talked about preserving his seven hundred hectares of rainforest, he sounded like a man trying to prove to himself what kind of person he was.

The soy field just in front of us, I realized, was about the size of a single hectare. In general, I have no sense for what an acre is, much less a hectare. But I had looked it up, and here was one in front of me: a piece of land about the length of a soccer field on each side.

"Isn't that a huge piece of land?" Rick said. "And then there's seven hundred of those back there in that forest." Rick's piece of rainforest suddenly seemed incredibly massive—an entire world.

"I still have a hard time even believing that I own that piece of

property . . . This shouldn't even belong to a human being!" He laughed.
"To have that kind of ability or power or whatever it is, or . . . "

"It's not like owning a car, or a house," I said. "It's like owning a little
universe that you're inside."

He nodded. "And it's been there since, oh, you know. It's evolved for
millions of years . . . And what gives me the right to just be born and all of a
sudden it's mine? I'm like a speck on the Earth. I've only been here for like,
just a grain of sand in time, and all of a sudden I've got this ability to just
erase something that took . . . "

Rick shook his head and looked at the darkness on the other side of the
hectare, where his rainforest began.

"All of a sudden I'm here," he said. "And it's like, I'm the guy holding
the bag."

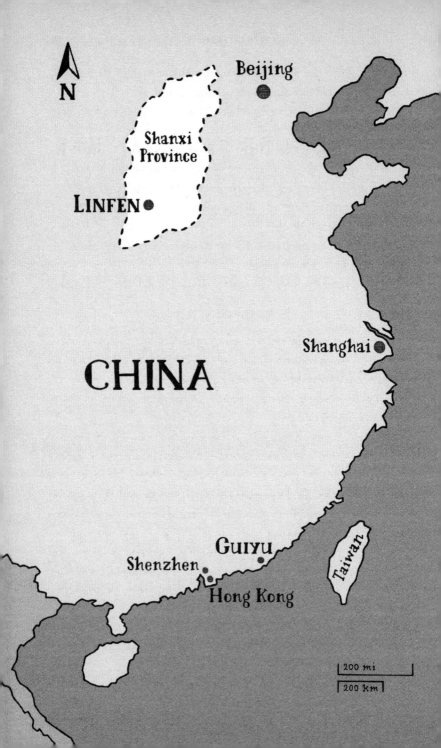

SIX

IN SEARCH OF SAD COAL MAN

Guiyu.

The smell of burning solder. Capacitors underfoot. Shattered components spilling from beneath a closed gate. Cellphone faceplates in heaps three feet tall, leaves raked up in autumn. We turn a corner. Ten-foot-tall drifts of gray computer plastic lie waiting to be sorted and recycled, like dirty snow dumped by a plow.

Old keyboards stacked on pallets, cube on cube, bales of electronic cotton. A warehouse of keyboards, a soccer field's worth of keyboards. A four-foot stack of identical keyboards, grimy and half crushed. I recognize the model; I used to own one. Has my old keyboard come through this place, for its keys to be ripped off, its metal extracted, its plastic melted down? I pry a key off and put it in my pocket.

A team of men shovel hay from the bed of a large truck, tossing it over the side into a heap. The timeless gesture of bodies shoveling hay, but it's not hay. They're shoveling circuit boards. Naked and green, the clattering square fronds pile up by the wheel of the truck.

Women toss piles of scrap aluminum into the air with shallow baskets, separating the wheat from the chaff. With broad, circular sieves, a family shakes out resistors and capacitors of different sizes. Did they come here from the countryside? Did they use these tools on the farm?

We did, said Mr. Han. We would use them for corn, back on the farm in Sichuan. Now, though, farms use machines to process the corn, and we just use sieves like these for sorting components.

Mr. Han had his own business. He and his wife had both grown up on farms in the Chinese province of Sichuan, to the northwest. They had met while working in an electronics recycling workshop here in Guiyu, near the southeast coast, and after marrying, they had opened a workshop of their own. They specialized in motherboards—the central circuit boards of personal computers. Mr. Han bought them in large bales three feet on a side, imported from overseas, likely North America.

Together with his wife and her sister and his wife's sister's husband, he sorted and processed every component of the motherboards. They cut out the valuable CPU chips for resale, pried off recyclable plastic, melted down and collected the solder that attached the components to the motherboard, sorted the components into sacks, and sent the cleaned motherboards off to have their gold extracted.

Theirs was one of thousands of similar workshops in town. Guiyu's entire economy is based on tearing apart old electronics and reselling the components and raw materials. Walk the streets and you will see building after building with a workshop at ground level and family quarters on the upper floors.

It's a dirty business. Computers are full of all kinds of things that are bad for you—things other than the Internet—and when you tear them apart, or melt them down, or saw them into pieces, a portion of those toxic substances is released. In a place like Guiyu, with what I'll call relaxed workplace standards, you end up with workshops full of lead dust and other heavy metals and clouds of who the hell knows what floating through the streets. The water is laced with PCBs and PBDEs and other hazardous acronyms. The air, the water, the dust—in Guiyu it comes with promises of cancer, nerve damage, and poisoned childhood development.

Exporting toxic waste across borders, especially to developing countries, is supposed to be illegal. The Basel Convention, the treaty that outlaws it, was already nearly twenty years old by the time I visited Guiyu, in 2011. In the case of electronic waste, though, the convention is easy to circumvent. As the green-electronics coordinator at the ever-present Greenpeace has said, "the common way exporters get round existing regulations is to relabel e-waste as second-hand goods for recycling."

Of course, it *is* recycling. Which is another thing, along with the town's curiously agricultural character, that complicates any appreciation of a place like Guiyu. But whether you consider it a toxic hellhole or a paragon of recycling and resourcefulness, the rivers of junked electronics flow in.

That it makes economic sense to ship the stuff halfway around the world for recycling is explained first by the low cost of labor here. But you must also consider the volume of empty shipping containers returning to China. Incredible amounts of manufactured goods are sent from China to the West in shipping containers, and since the conveyor belt must run both ways, sending freight back is cheap. The result is that we don't really buy our electronics from China after all. We just rent them and then send them back to be torn apart.

India and certain African countries, including Ghana and Nigeria, also get in on the game, but China is the e-waste importer par excellence, and Guiyu is the industry's crown jewel. Guiyu is so famous for its commitment to electronic waste that it has become a mecca for journalists interested in the topic—which some people here don't like. In 2008, a crew from *60 Minutes* was attacked while filming a television report in Guiyu. Shady businesspeople don't want their dangerous, quasi-legal industry exposed. But maybe there was an element of local pride as well. If my town were world famous as a warren of poisonous bottom-feeding, I'd probably be pissed off, too, when people wandered into my workshop with cameras. Whatever the source of the bad vibes, Guiyu sounded unfriendly. I had heard stories of journalists being screamed at, chased, pelted with bricks.

Guiyu isn't the only weirdly specialized place in Guangdong Province.

Only two hundred miles down the coast is the "special economic zone" that is the city of Shenzhen, one of the most concentrated areas of electronics manufacturing in the world. (It was to companies in Shenzhen, Mr. Han said, that he sold his recycled components.) Shenzhen is home, for instance, to the famous "Foxconn City," the giant complex where iPhones and a million other things are built.

From waste recycling to questionable industrial processes to simple carbon emissions, Guangdong is a land to which we outsource not only our manufacturing but also our pollution. The environmental reporter Jonathan Watts put it best, in his book *When a Million Chinese Jump*: "This is where the developed world dodges its own rules."

Then you have Gurao, the Bra Town of Guangdong, just up the road from Guiyu. Passing through Gurao on the bus, I saw billboard after billboard of semi-nude lingerie models. Colossal women in bras looked down from the facades of factory buildings. One lounged next to a violin. Nearly all the models were Western; full hectares of white flesh went by. A rippling male abdomen crowned a pair of tumescent briefs—the work of the Guangdong Puning Unique and Joy Clothing Co. Hanging from the streetlights, where another town might fly banners celebrating a holiday or a music festival, there were pennants with more white people in their undies. The children of Gurao must grow up thinking that without their city to stand in the breach, the Westerners of the world would go completely naked.

I saw bras, but Cecily smelled a story. Cecily was my fixer and translator, a young Chinese reporter whom I had hired in Beijing. She was intrigued by Bra Town. She wanted us to pose as entrepreneurs interested in importing bras to the United States. That way, she thought, we might get a look inside one of those factories.

That we would consider working undercover to get an inside look at the presumably legal underwear industry was symbolic of a broader problem: in China, I was not supposed to be a journalist. My tiresome habit of telling myself I wasn't one anyway made no difference. Several people, professional reporters with years of experience in China, had advised me to travel on a tourist visa, not to be open about my agenda as a writer, and not to do anything that could draw the attention of local authority figures or media, not to mention those unfriendly guys with the bricks in Guiyu.

As for Cecily, she said I should specify that she was a tourist guide and wasn't doing any journalistic work. (She was a tourist guide. She did no journalistic work.) And there were larger things afoot. The Chinese government had been spooked by the revolutions of the Arab Spring; before I left the country, it would begin a crackdown that included police intimidation of foreign journalists, and even some violence.

As we rode the bus to Guiyu, Cecily described what she called, with irony, the "fun game" played between Chinese journalists and their government. Reporters would play cat-and-mouse, testing just how far they could go. Items that might be barred from a general interest newspaper, for instance, might be allowed in a specialty magazine. Censors were most active on the Internet at specific times of day, so posting a piece online at the right moment could allow it to find its audience for a few hours—even if the publisher would then cooperate with the censors in taking it down. The government, she said, was "insane about journalism."

At the moment, our need was to concoct a cover story under which I could both snoop around and plausibly deny that I was a reporter. We soon realized, though, that while my credibility as a reporter was poor, my credibility as everything else was even worse.

"Maybe you are a professor from a university," Cecily said, evaluating me with a sidelong glance.

Was she serious? Were there really professors out there quite this haggard and blotchy-faced? This badly dressed?

"Maybe you want to open a shop," she said. "A businessman of some kind."

Yeah, right, I thought. I'm one of those businessmen who look like they've forgotten to shave for twenty years.

I suggested *artist* instead. Everyone knows that artists can be plug-ugly and sullen, but with a strong undercurrent of narcissism. I was perfect for the role. And it would explain all the photographs I wanted to take.

Cecily was skeptical. She didn't think the idea of an artist seeking out polluted places would translate. Maybe she had a point. Besides, Edward Burtynsky got there first.

Then it struck me. We didn't need a cover. We just needed a *joke*. Humor was the universal language. We would tell everyone that I had thrown my old cellphone away by mistake and had come to Guiyu to retrieve it.

Cecily laughed. "I think artist is better."

\\\\\\\\\\

Nobody threw bricks. Instead, we found ourselves under a heavy tea barrage. Through blind luck, we had found the Han family and were now enduring the withering assault of their generosity and good spirits.

After walking down the Guiyu streets for an uncomfortable duration—uncomfortable for the way we stuck out, for the way people stopped what they were doing to watch us and possibly ready their bricks—we came upon Mr. Han sitting in the doorway of his workshop. He was youngish, perhaps in his early thirties, and had a friendly face. His forehead and hair were powdered with dust. He had been using a small circular saw to cut CPUs out of a select stack of motherboards. In Chinese, Cecily asked if we could see his workshop.

Like their neighbors, the Hans lived on the upper floors of their building, reserving the ground floor for a garage-like workroom. One corner of the workshop was a sitting room with a teakettle and a computer; the rest was filled with piles of motherboards, shelves of CPUs, and large grain sacks filled with sorted resistors and capacitors. We sat and drank tiny cups of tea by the half dozen while the family's tiny, eight-year-old son made a racket throwing circuit boards around in the back of the workshop.

Mrs. Han wanted to know why Cecily, in her late twenties, wasn't married, and whether I was married, and whether two single people traveling together were perhaps soon to be married to each other, and finally, once again, whether I was married.

"Is he married?" she asked, looking at me with cautious amusement, as though I were a zebra.

I said I was not. Married. I didn't elaborate. I was in fact more than unmarried. I was newly alone, and homeless. After getting back from Brazil, I had moved out of the Doctor's place. Now, when not in Guiyu, I resided on an air mattress on Adam's living room floor, where I spent my nights praying to be hit by an asteroid.

We began the business of lying to our new friends. Cecily and I had not agreed on a cover story in the end, but the Hans quite naturally wanted to know what had brought me to Guiyu, and to their workshop. Improvising, Cecily threw out several stories in quick succession, no doubt creating some confusion as to exactly what an artist/university researcher/entrepreneur was. I told Cecily I was worried they wouldn't buy it.

It doesn't matter, she said. We just have to tell them something.

In the meantime, I had realized that the little tyke in back wasn't thrashing around just for fun. He was working. I told Mr. Han that I'd be happy to relieve his son for a while. I was a hard worker, I said, a claim that proved wildly hilarious to the entire family. When the laughter died down, I was still looking expectant.

Is he serious? Mr. Han asked.

I think he is, Cecily told him.

Mr. Han shrugged. Well, sure. Lang can show him how to do it.

And that is how I began my career in electronics recycling, in the employ of an eight-year-old firebrand called Lang. Our task was to pull the recyclable plastic off the circuit boards, which were piled against the wall in a mound almost as tall as I was. We sat at the foot of the mountain on tiny plastic stools, causing little avalanches each time we grabbed a new board.

Most of the recyclable plastic in a computer's motherboard, I'll have you know, is in the slots where sound cards and the like are plugged in. With the use of a screwdriver-size crowbar and a pair of pliers, these narrow rectangles of plastic can, if you are Lang, be popped off the board with a few flicks of the wrist. Lang also had a preternatural ability to move boards around with his feet, leaving his hands free for uninterrupted hammering and prying. He was wearing a pair of fuzzy brown dog slippers with floppy ears, which created the illusion that he was being helped in his work by a pair of supremely well-coordinated puppies. He was a machine. In the time it took me to evict a single battered hunk of plastic, Lang might have gone through three entire boards, plastic flying from his every touch, the boards spinning underneath the jumping ears of his little doggies.

I held up a newly won gobbet of plastic. "Check *that* out," I said with pride.

Lang smacked his forehead. "Bu yao!" he cried, and snatched my board away.

"Cecily!" I shouted to the sitting room. "What does 'buyao' mean?"

"It means 'don't want,'" she said.

It turns out there is no better way to learn a few useful words of Chinese than by taking part in a little child labor. In addition to *bu yao*, I learned *yao*, meaning "want," and *hao le*, meaning "done," roughly. Like this, Lang and I established a system of communication, and I began to learn which bits of plastic were worth the prying and which weren't. Some, I believe, had metal inside them, and so were no good for melting down.

Over the course of several hours, Lang's excitement at getting to boss around an adult veered into delight at what was becoming an effective collaboration. Soon, when he would go to get a smoke for his uncle, he would get one for me as well, leaving me with a lit cigarette in my mouth before I could even think of saying *bu yao*.

The smoke stung my eyes as I worked, making me glad that we were not baking circuit boards instead. That task was done in the covered entry space between the workshop and the street, and was a job the Hans didn't do themselves. They reserved it for their lone employee, who sat in front of a hot plate that held a shimmering pool of molten solder. With a pair of needle-nose pliers, he would pick up a circuit board and float it on the silvery pool of solder. As the solder holding the components on the board melted, acrid fumes rose into a homemade fume hood, which drew them into a chimney and vented them onto the street. This is why the streets of Guiyu smell of cooking circuits. Nearly every building has one of these smokestacks.

After frying for fifteen or twenty seconds, the circuit board's connections would melt. The worker would pick up the board with his pliers, invert it, and smack it violently on a hunk of concrete to the right of the stove. The components would fly off (along with a spatter of tin and lead, depending on the solder) and go tumbling into an ever-growing pile. He would then toss the board into a heap of newly naked circuit boards.

There was gold in those boards. Printed circuit boards use copper for their circuits, but the copper must be protected from corrosion with some kind of coating or plating, often in the form of a microscopically thin layer of alloyed gold. It takes a lot of circuit boards to accumulate a significant amount of gold, but a lot of circuit boards is exactly what Guiyu has. Once Mr. Han had accumulated a sufficient batch, he would give the boards to a contractor to extract the gold. This was the dirtiest part of the entire process. I had heard tales of acid baths and toxic bonfires. Naturally, I wanted to see it for myself.

Don't, said Mr. Han. Don't try to find those people. They operate illegally, and they're very suspicious. You could get in trouble. Please don't try to find them.

Not that I had time anyway. I was focused on my work, on improving my turnaround time for each motherboard. Brand names cowered under my crowbar: Intel, Acer, Foxconn, Pentium, Philips, Virtex, Blitzen. Each time I had a CPU to unplug from a board, Lang would hold up the collection bucket for me, and I would shoot a three-pointer, and he would smile like we had won the championship.

A drag on my cigarette and I'd pull over another board to wreck out the plastic, pausing to point when I wasn't sure.

"Yao?" I would ask.

"BU YAO!" Lang would scream.

"BU YAO!" I would scream back.

And then, if I thought I was done, I would ask, "Hao le?"

"Hao le," Lang would say, sounding almost philosophical. Then, with a look of what I hoped was respect, or at least camaraderie, he'd pause his helper-dogs and slide another board in front of me, the little slave driver.

\\\\\\\\\\\

A storefront with a small glass case full of integrated circuits. It was a tiny shop, one room, run by two young brothers. Three feet behind the display case was a bunk bed. They lived in the shop. To the left was a table arrayed with a hundred small cups for sorting their wares. One brother, wearing a red pleather jacket and a striped button-down shirt, watched with cautious amusement as I took pictures of his display case.

I wanted to buy a small baggie of chips as a souvenir. He was confused. What did I want it for? The kind of chip I should buy depended on the intended use. When he finally understood, he refused to let me buy one, insisting that I accept it as a gift.

On a busy market street, a nail salon with six young women in black tights and high-heel boots. All the young or youngish women in Guiyu dress this way. They chattered as they bent over their work. It was of course not a nail salon but a circuit shop. Each woman held a handful of chips. Using tweezers, they would pick up a single chip and dip each of its two rows of contacts into a pool of molten solder on a shared hot plate, working with the speed and economy of motion that comes from day upon day of precise repetition.

We asked if we could take a picture of them working. They tittered. One of them, in the second it took her to pick up her next handful of chips, waved her free hand in front of her face and smiled. Please don't.

We wandered the streets, passing over small canals choked with trash. But trash-choked waterways are like sunsets. They're great to look at, but they may not mean that much. More interesting are the many smells present in Guiyu, the many shades of water and air that complement the clouds of fried circuitry. At the river, drifting stains and a reek of sewage. Near the bus station, a generalized fetid-toxic smell hanging over a canal by the road. On the bridge, an inky stink of exhaust coming from a passing tractor-tricycle. I watched with some dismay as the choking plume approached us. But then, as the driver passed by, he throttled down for a moment, sparing us the worst. Even in Guiyu, courtesy lived.

Through a back alley we came upon a crew working through pallets of Motorola Broadband Media Centers—cable boxes. A man had stacked about fifty of them along one side of the work area, forming a wall of identical metal boxes, and was going from one to the next with a screw gun, unscrewing the same four screws on each. Behind him his coworkers made tidy piles of tops, of sides, of brackets, of LCD screens that trailed ribbon cables—a tangle of color on a dreary afternoon.

Trucks belched along with loads of semiconductors. A motorcycle cart passed us carrying a pile of strange, green objects. With a start, I saw they were cabbages.

We paused by a truck, its bed loaded high with bulging sacks. The corners of cleaned circuit boards peeked out from the bags. Raw material, about to be hauled off to the mysterious gold extractors, wherever they were. The men loading the truck smiled and asked where I was from.

Mei guo, we said. America. What's in the truck?

They smiled a little less. Cardboard, they said. Paper. For recycling. And they got in their truck and left.

A gaggle of teenagers waylaid us and led us on a short tour, to a community center, where teachers tried to control a restive mob of music students. We were a sensation. For a moment I knew the life of a rock star, reducing his fans to convulsions with a single moment of eye contact.

Our abductors took us to a nearby temple. This is our temple, they said. We walked through crumbling, ornate rooms overseen by a platoon of deities and demigods.

You should pray here, they said. To this god. Make a wish as you kneel and bow, with your hands together. So I did it. But I couldn't decide whether to wish for peace or for love.

\\\\\\\\\\\

The secretary of the Guiyu business association—or whatever it was—met us in the evening at the Six Star Coffee Shop in Shantou, the large coastal city where we were staying. The Six Star had two levels, every seat a sofa, including several sofa-like things that hung from the ceiling on cords, porch-swing style. It was a place where the wealthy and cosmopolitan of Shantou could gather to feel wealthy and cosmopolitan. The menu was broad and evocative, with helpful descriptions in English. I wavered over "Irish Coffee—Emotional, romantic, and mysterious" before settling on a latte, because "the latte's mellowness with the Hazel's aroma, Special flavor. Men's favorite."

The secretary wore a puffy red jacket and stylish eyeglasses. She had brought along her teenage daughter, a docile, wide-eyed girl who ordered an absurdly large pink drink. It exploded with fluorescent straws and a large wedge of fruit cut into an artful splay that evoked a breaching humpback. Her mother ordered a pot of fruity tea.

Cecily had chosen "university researcher" as my cover this evening. To my amazement the secretary accepted it without a blink.

We want to improve the environment, she said. But as she had only been on the job for a couple of months, she didn't know much about the industry she represented. Maybe that was the point. We were originally supposed to meet the associate director, until he decided otherwise and foisted the secretary on us. She punted question after question by saying she would send us some informational materials put together by the association. (She never did.)

Since she had brought up the environment, though, I felt comfortable asking her about emissions and workshop conditions. She said emissions from burning circuit boards were the main environmental problem.

I doubted it. The Hans' workshop, for instance, although host to a warm and supportive family atmosphere, was almost certainly powdered with lead, tin, and antimony dust, not to mention other toxins from all the sawing and board frying. So when little Lang and his sister came home from school to help out in the workshop, they were not just taking part in the family business. They were most likely being poisoned. In this, they were representative of both Guiyu and a wider phenomenon. In its pursuit of unfettered economic results, China has allowed widespread lead poisoning. This is especially dangerous to children, whose nervous system and mental health can be permanently damaged. "In more developed nations," the *New York Times* said in June 2011, "a pattern of lead poisoning like China's would most likely be deemed a public-health emergency."

The secretary told us that the government had recently started taking the environmental problem seriously. And the business association was trying to attract investors and start partnerships to develop new technology to do the work more cleanly. Again, I doubted it. The problem wasn't technology. It was that to be economically viable, the e-waste industry operated unsafely, and was allowed to.

The secretary asked me a question. Did I have ideas for new technology?

Me? I may have misunderstood Cecily's translation. The secretary was asking *me* for ideas of how Guiyu could do its business more cleanly? Or for institutional contacts? What should I say?

I smiled blandly and nodded, in a way that conveyed neither comprehension nor intelligence.

"Not off the top of my head," I said.

It's okay that journalists have come to expose the problems, the secretary said, pouring some more fruit tea. But it's more important to find solutions than to criticize.

\\\\\\\\\

We made another visit to the Han family the next day. We wanted to thank them and to say goodbye. Also, when you're in a strange town where you don't know anybody, it's nice to go someplace where people will smile and offer you tea and cookies.

You're sure he's not a journalist, Mr. Han asked Cecily.

No, no, she said.

By now I had fully developed the knot of guilt in my stomach. Mr. Han wasn't stupid, even though, with our cockeyed cover stories, we may have treated him like he was.

Today, adding to my shame, he offered us lunch. Upstairs, around a low table in the kitchen, we ate meat and vegetables in the Sichuan style, and a spicy dish of preserved black beans from the family farm, where their parents and extended families still lived. The Hans sent them money regularly. That was why they had come to Guiyu in the first place; there wasn't enough work where they came from. They'd been here for fifteen years. The locals, they said, still treated them like outsiders.

Back downstairs, we had another three dozen small cups of tea. Mr. Han sat in front of the computer, paused the movie that was playing, and checked the commodity prices. Figures filled the screen. It was important for him to know the current price of gold and other materials so he didn't get ripped off by his buyers. His computer also stored the video feeds from the security cameras in his workshop. With a few clicks, he brought up a high-angle shot of Lang and me raining grief on circuit boards.

Mrs. Han wants to know again why you're not married, Cecily said.

They had asked a dozen times. They couldn't have known that I spent most of my free time asking myself the same thing. I realized, though, that this was an opportunity for me to answer at least one of their questions honestly.

"Tell them that I was going to get married, but the woman changed her mind," I said to Cecily.

She translated.

They say that's terrible, Cecily said. That it's really embarrassing. But that I shouldn't tell you they said so.

"Do they have any advice on how to find a good wife?" I asked.

Mr. Han nodded. Choose someone who loves you, he said. It doesn't matter if you love her. Just make sure she loves you.

I couldn't decide if this was horrible advice or profound. "Shouldn't we both love each other?" I asked.

Choose someone who loves you and who takes care of you, advised Mrs. Han. Don't just choose someone who you love. And if there are things you don't like about the person, you'll come to see past those things and love her eventually.

They told us their love story. Mr. Han had pestered his wife-to-be to give him rides to work on her scooter. They had written a long series of love letters. Mrs. Han said she still had the letters he had sent her.

Cecily asked Mr. Han if he still had the letters Mrs. Han had sent him in reply. He shook his head, and his wife rolled her eyes. Men aren't romantic, she said. They don't keep that stuff.

Mr. Han was smiling. He pointed at his chest. I keep them in here, he said. I keep them in here. And everybody laughed.

We stood to go, waving to Mr. Han's brother-in-law, who was working his way through the last few boards of Lang's mountain from the other day. Lang and his sister were at school, because in Guiyu that's what *jawas* do on weekdays.

In the foyer, I drew a lungful of frying circuit board. It reminded me of something Mr. Han had told me earlier. I had asked him if he thought the work was unhealthy for him and his family.

We know it's a dirty business, he had said. We know it's a health risk. You have to give something to get something.

As we left, he was standing in the foyer of the workshop, contemplating two bales of motherboards that had just arrived. The next batch. He had slashed them open at the side, spilling fresh, untouched circuitry onto the floor.

\\\\\\\\\\

At the Beijing airport, the sun peered through a thick scrim of haze. Several years before, in preparation for the Olympics, the Chinese government had gone to extreme lengths to reduce the city's famous smog. Anything for a coming-out party. If this was reduced smog, though, it was still pretty impressive. I had noticed the haze days earlier as well, on our way through the airport to Guiyu.

"Is that the famous Beijing haze?" I'd asked Cecily.

She looked out the window. "I think it's just because it's going to snow later today," she said. "The forecast is for snow."

Pre-snow haze?

"I think so," she said.

"No, Cecily," I said, laying down some ground rules. "It's pollution, okay?"

Now, on our way back, she abandoned the snow excuse. Instead, she mentioned that fog was in the forecast.

"There has been fog for three days," she said.

"Smog," I said.

"Fog."

"Smog."

"Fog."

She was an uncompromising negotiator. But later in the evening, after we retired to our hotel rooms, she sent me a text message to tell me that, on television, the news was that it had been the most polluted day of the year so far. I win, Cecily.

\\\\\\\\\\\

Night. Beside the highway, the squat, flaring glow of a refinery floated by, bladerunner-like in the haze. We watched from the pitch-darkness of Liu's cab.

The most polluted city in the world. The beams of oncoming headlights writhed in the heavy air. We were driving to Linfen. Through the city's outskirts and onto a broad multilane highway, an empty avenue of street-lights, the smog unbelievably thick. We passed a carved sign: WELCOME TO

LINFEN. With perfect timing, a truck piled high with coal came onto the road in front of us. A chunk of coal skittered loose and obliterated itself against the roadway, joining the stains that marked the passage of previous coal trucks.

Linfen is a coal town, and legendarily dirty. In fact, hardly anybody outside China has ever heard of the place, unless they've heard of its pollution. That was the only reason I had heard of it, and the only reason you're hearing about it now. Linfen sits at the heart of China's coal country, in Shanxi Province. But to visit Linfen is not merely to travel to another time, to remember how industry used to dominate the landscape around American and European cities. Linfen is also a convenient symbol of what China is doing to the global environment—the same thing we've been doing for a hundred years.

Let's run some numbers. China's consumption of coal doubled in the decade leading up to 2010, to more than three billion tons. That's nearly half the world's annual supply. About three-quarters of China's electricity comes from coal, and as the country's electrical needs have skyrocketed, so has coal-fired power generation. China is not only far and away the world's biggest consumer of coal but is also its biggest consumer of energy, and its biggest emitter of carbon dioxide. And even though China is fast becoming a world leader in renewable energy, like wind and solar, it is coal that has powered the nation's precipitous rise.

This matters. The coal gets burned over there, but the carbon dioxide goes everywhere. So if, by some miracle, the West manages to stop screwing over the global climate—well, we probably won't. Regardless, China has picked up where we haven't left off.

The most polluted city in the world. We were downtown now. There was dust everywhere, thick on the cars, thick in the air, coating the buildings. *Finally*, I thought. Someplace really grim. A polluted place that isn't nice. A place I can point to and say, *Yes. It's even worse than you imagine*.

Cars swam past in the murk as we parked and headed into the hotel. In the lobby, the staff seemed to have lost our reservation, seemed in fact surprised that anyone would want to stay in their hotel for an entire night. A dwarf stood by the desk, looking me up and down with a sneer of disbelief.

I shuffled upstairs to my room, past the hall attendants, who insisted on taking my key and opening the room. It seemed less a point of hospitality than a security procedure. My room smelled faintly of cigarette smoke and urine, and I went to sleep grateful, having at last found what I came for.

\\\\\\\\\\

The next day, though, the spell was broken.

Preferring something less redolent of gambling and low-rent organized crime, we changed quarters, moving to the Honglou, a nice hotel near the university and in sight of Linfen's drum tower. After dropping our bags, we went to check out the city.

If you're counting drum towers, the one in Linfen is supposed to be the second tallest in China, at 150-some feet. At the base of the tower, a worker was sweeping the sidewalk, pushing a carpet of beige powder as he went. I noticed again that everything in Linfen was dusty. A brown film coated cars that had been parked outside for even a single day. In the lobby of the Honglou, a woman had pushed a dust mop back and forth over the wide marble floor, Sisyphean and smooth.

We climbed the tower's dusty stairs. Inside, we stared up at the ornate wooden vault of the ceiling. Drum towers and bell towers used to be important features of Chinese cities, timepieces to mark the day's passage. But that's all over now. Besides, when exactly would you drum the sunset in Linfen? When the sun disappears behind the smog? Or sometime later, when you assume it has reached the horizon?

From the balcony, we looked onto the hue and drone of the traffic circle that surrounded the tower. To the south, down the crowded boulevard of Drum Tower Street, we could see only a few blocks before the traffic faded away into the haze. It was like a thick mist lay on the city—but there was nothing misty in this mist, nothing damp or fresh.

At this moment, though, something began to dawn on me. I was having that feeling. That good feeling. The sensation of having woken up in an interesting new place. *Oh, no,* I thought. *Not again.*

Was Linfen really all that bad? True, its smog was the smoggiest smog I had ever seen. Smog to irritate your throat. Smog to keep you coughing through the night.

Still. I pointed my camera down Drum Tower Street. If I zoomed in all the way and took a photo where the buildings dissolved into the murk, Linfen appeared oppressive, unbearable. But if I zoomed all the way out, Linfen looked like . . . just another place.

Later, I showed the zoomed-out photo to my friend James, to show him how, at a visceral level, Linfen wasn't so horrifying. He looked at me archly and said that, to him, it still looked pretty terrible. His amateur meteorologist side kicked in: he estimated the visibility in the photo at a quarter mile. The same as in a heavy snowstorm.

So don't let me tell you it's not bad. It's bad. *It's really bad.* Chronic respiratory disease and even lung cancer must stalk the city's boulevards and alleyways. Schoolchildren surely contend with lungs seized by asthma. And doubtless, Linfen is symbolic ground zero for what the human race is about, these days. But when I looked down on the city from the drum tower, I saw not only smog but also cars and buses, and the KFC, and people going about their lives.

I put it out of my mind. We went down the stairs and crossed the street to check out the large civic plaza that faced the tower. Drum Tower Square, as I choose to call it, was festooned with decorations for Spring Festival. *Festooned* is the only word. Large mutant rabbits made of wire and fabric loomed over us. It was the Year of the Rabbit, and although Spring Festival—that's what they call Chinese New Year in China—had already ended, that didn't save us from being leered at by cartoon bunny rabbits everywhere we went.

The main problem with the plaza was its heartwarming display of healthy civic life. People gathered here and there in small crowds, singing old Communist anthems with obvious nostalgia. Passersby came together in circles around street musicians. In the back of the plaza, an ad hoc dance hall had been set up, complete with amplified music. Couples twirled through an unorthodox rumba. One pair glided across the stones of the plaza with eerie smoothness, the woman's long black hair swinging over the purple velvet of her overcoat.

The dance music, too, was an old propaganda song, Cecily told me. *The Communist Party saved the people*, went the lyrics. *The dearest people of all are the communist soldiers.*

"People don't really take this music seriously anymore," she said.

At its southern edge, the improvised ballroom came up against another dance area, where a rank of about a hundred people, mostly elderly, were proposing a variation on the electric slide. They beamed with carefree amusement as they danced. Who were these happy citizens?

And above all, where was Sad Coal Man?

Who is Sad Coal Man? Search Google Images for "Linfen," and you'll see him. Nobody gives his name, of course, so I just think of him as Sad Coal Man, and if there is an iconic image of Linfen, he is it. (The silver medal goes to Sad Coal Man's older brother, Man on Bike with Face Mask.)

Sad Coal Man's lot is to stand forlornly by the side of the road, forever staring into the distance over our left shoulder. Sad Coal Man is young and wears a dirty brown jacket over a dirty brown sweater, with a dirty black shirt underneath. Sad Coal Man's face and neck are covered with coal dust and his brow is furrowed. When reproduced at a small image size, he looks like he's squinting, almost in pain. Larger versions reveal more subtle emotions. His eyes are clouded not with pain but uncertainty, with doubt for the future. Sad Coal Man is so sad he looks like he might cry. But he can't. His heart has been hardened beyond tears by a lifetime lived in the world's most polluted city. Sad Coal Man also needs a haircut.

Never mind the blue sky in the corner of the photograph, over his shoulder. With Sad Coal Man as evidence, you can draw only one conclusion: Linfen is a hellhole, a place bereft of human dignity, where people don't even know how to wash, because there's no point. His expression and appearance are calibrated to bring out our condescension. *It's so terrible they have to live that way.*

When I look at him now, though, I see something else in his face. Awkwardness. Someone has told him, *Stand here. We're going to take a picture of you. Don't look at the camera.* I'm willing to bet that Sad Coal Man wasn't thinking about the plight of Linfen when they took his picture. He was probably thinking, *I wish they'd let me wash my face first.*

But Sad Coal Man was nowhere to be seen in Drum Tower Square. Maybe he was up in the mountains, mining. Maybe we'd find him later, and ask him what he was thinking in that picture, and whether he was friends with the Crying Indian from those anti-littering ads of the 1970s.

The square had more to show us. On the other side of the semi-electric slide, people were playing hacky sack. In this part of the world they use a weighted, feathery shuttlecock, but the moves are the same: the inside kick, the outside kick, the chest check, the behind-the-back. The only difference is that in Linfen—perhaps in all of China, I don't know—hacky sack is not just a game for young men, but for people of all ages. Best of all were the grandmothers hacking it up like they were between classes at Hampshire College.

Nudging toys and rabbit-shaped balloons out of the way, we ducked in front of a row of vendors. There was writing on the ground. Half a dozen men were practicing calligraphy, using long brushes to paint water on the stones of the plaza.

That was the last straw. The civic charm offensive was complete. To grow old within walking distance of Drum Tower Square seemed like a blessing, if you had the lungs for it. Here, in the smog capital of the universe, I was reminded that there was more than one kind of health.

Sometimes I despair at the prospect of growing old in my own country. In the United States, seniors are supposed to keep to the house, or at least stick to the park benches. You don't exactly see them playing Frisbee in Central Park. In Linfen, though, citizens old and young come to exercise in the public square, and sing old songs, and play hacky sack. They dance, they slide electrically, they watch their kids or grandkids ride plastic tricycles around like lunatics. They write poems in water on the flagstones, and watch them evaporate. This place was pretty great.

Don't worry. I'm not debunking anything. We're still ruining the world, and Linfen is still polluted as hell. The reason I find myself beating the same thematic horse on every continent isn't that the polluted places of the world aren't polluted. It's that I love them. I love the ruined places for all the ways they aren't ruined. Does somebody live there? Does somebody work there? Does somebody miss it when they leave? Those places are still just places. But when we read horror stories about them at home in our cozy green armchairs, we turn them into something else, into stages on which our worst fears can play out.

We also hold up these poster children—Linfen, Port Arthur, Chernobyl—to tell ourselves that the problems are *over there*. And we'd like to keep it that way. We'd like to keep a tidy bubble for ourselves, and draw a line around some trees, and declare *no farther*. That *here*, at least, inside *this* boundary, nature survives. As long as there is Yellowstone, we'll have a little something for what ails us. What a joke. So much of our environmental consciousness is just aesthetics, a simple idea of what counts as beautiful. But that love of beauty has a cost. It becomes a force for disengagement. Linfen is too foul to care about. Port Arthur is too gross.

So I love the ruined places. And sure, I love the pure ones, too. But I hate the idea that there's any difference. And I wish more people thought gross was beautiful. Because if it isn't, then I'm not sure why we should care about a world with so much grossness in it.

One calligrapher finished painting a broad grid of beautifully rendered characters, and several of his fellows began a jocular critique of his work. An aging man with a dark green jacket and a bad comb-over saw us watching, and stepped forward.

His name was Mr. Ma, and he wanted to know if I could understand the conversation we were listening to. Cecily told him I couldn't.

But foreigners are smarter than Chinese, he told us, not even half joking. He had heard a foreigner speak Chinese once, and had concluded that it must be very easy for foreigners to learn it. He thought I must understand it, too.

Disbelief that I didn't understand Chinese had been a running theme. In Guiyu, Mrs. Han had asked Cecily about it more than once. He can't understand us? *At all?*

A retired prison guard, Mr. Ma had lived in the Linfen area all his life. The city had expanded over the years, he said, but it hadn't changed much. Had I noticed the air?

I had.

It's haze and coal, he said.

Yes indeed, I said.

He addressed Cecily. Be open-minded about dating foreigners, he told her. It's okay for Chinese and Americans to marry now.

Cecily rolled her eyes. I think. I couldn't really tell, as I was busy with my own eye roll. To his credit, though, Mr. Ma also told Cecily that she had done right to focus on her career and education.

Take care of him, he told her, as we parted ways. He's a guest in our country.

\\\\\\\\\

Linfen has a number of decent attractions besides the smog. From Drum Tower Square, you can take a nice run through the grounds of Shanxi Normal University and out to the riverfront, which reminded me a bit of Hudson River Park, in Manhattan. There were no skate parks or beach volleyball courts, but there was a mini-golf emporium, closed for the winter.

Spreading east from the river is the Hua Gate area, a wide pedestrian arcade in the style of the National Mall, lined with temples and buildings. Of these, only the Yao Temple is supposedly original, the site of one of the earliest Chinese dynasties, dating back millennia. But it's hard to know what's real. Across the arcade from the Yao Temple is a scaled-down replica of the Forbidden City's famous Meridian Gate. The fakey vibe only increases down the sidewalk, where there's a replica Temple of Heaven, distinguishable from the Beijing original not only because it is much smaller but because it has a haunted house inside.

From a cart, we bought two *rou jia mo*, large flat biscuits stuffed with pork and onions and peppers. In the cold air of winter, devouring my sandwich, I decided it was the most delicious thing I had ever eaten.

Next to us was a large sign with a picture of the Hua Gate, a sort of oversize Arc de Triomphe. ONE OF THE FIFTY MOST WORTHY PLACES FOR FOREIGNERS TO VISIT, the sign read. NATIONAL AAAA TOURISM SCENERY.

Good enough for me. We started toward the far end of the mall, passing carnival games and rides, hangers-on from a Spring Festival installation that gave the whole place a Coney Island feel. We stopped for a bit to explore a large, concrete relief map of China, its crumpled mountains reaching halfway to my knee. We stomped over the earth, leaping from one stone section of the mini–Great Wall to the next, clearing entire river systems at a stride. I happened upon Sichuan, the home province of the Han family, and from on high peered down on its rosy surface, the paint coming up in flakes.

At the far end of the boulevard was the Hua Gate. Only a few years old, it was the largest gate in the world, a man told us. But it could only generously be called a gate.

"If it's a gate, you should be able to drive through it, or something," Cecily said.

But the Hua Gate's purpose was not to be driven through. Instead it was some kind of Chinese national bicep for the flexing, the gaudy ornament of a nation newly confident of its dominion. It was a monumentally ornate, gate-like building, its vast doors closed off with walls of glass. Inside we walked across its marble floors and past its huge, colored pillars to find the stairs. Three stories up, under colored LED lights that gave the place a definite Vegas sheen, we encountered a Hall of Great Chinese, with thirty-two gilded statues of this emperor or that navigator or that inventor, all of them ancient, dating to an era somewhere between history and mythology.

In the center of the room was a translucent hemisphere with the out-lines of what looked like seven continents floating on its glass surface. It took me a while to realize that they were not the seven continents but rather seven different iterations of China, the outlines of seven different dynasties through the ages, now floating free across the globe, unimpeded by other land.

The next hall up hosted the statues of thirty-two famous Chinese women. They floated in the moody, blue-pink light. Cecily's eyes went from one to the next, wondering if someday there might be room for her.

Lying in the middle of the room, twenty feet tall if she had stood up, was the grandly naked figure of Nu Kua, the goddess who first created human beings. The humans she had created frolicked all around her: freaky little golden babies that looked to me like they were up to no good.

\\\\\\\\\\

On the city outskirts, we stopped so I could take some pictures of the bill-boards. There were advertisements for SUVs and sixteen-wheelers and even coal trucks. What had caught my eye, though, was a series of municipal ads. One had a picture of the drum tower under a suspiciously blue sky. The adjacent billboard showed an idyllic meadow scene, complete with fluttering doves. In the distance were city buildings; in the foreground, a ladybug perched on a photoshopped leaf. Above it all lorded a brilliant, shining sun. It's always nice to find propaganda that has an element of farce.

Overlaid on the picture was a message: LOVE LINFEN. PROTECT THE ENVIRONMENT. ESTABLISH THE IMAGE.

"Does he work for the environmental protection bureau?" the taxi driver asked.

"No," said Cecily. I was glad to hear her back away from that one.

The driver was a waggish young man who liked to talk. "I heard that foreign media declared Linfen the most polluted city. That was embarrassing," he said. "Is that why he's taking pictures of the ads? During the Olympics they shut down a lot of coal mines and polluting industries, so it's better now." They were no longer the number-one polluted city, he said.

Cecily asked him who had taken the lead spot. "I don't know," he said. "It doesn't matter. At least it's not us."

It was five years earlier that Linfen had first been declared the most polluted city in the world. The rankings were the work of the Blacksmith Institute, a New York nonprofit dedicated to fighting toxic pollution in developing countries. The group's website notes that, thanks to decades of environmental activism and legislation, "gross pollution" has been radically reduced as an acute problem in countries like the United States, but that in the developing world—out of sight and mind to most of us in the West—more than a hundred million people still face serious health effects from rampant industrial pollution and toxic waste. Blacksmith's mission is to attack this issue by pinpointing locations where concrete action could have major benefits for the health of a lot of people. The organization then provides grants and other support to local partners, who attack specific problems.

Blacksmith released its first public report in 2006, as part of its campaign to bring attention to such areas. Called *The World's Worst Polluted Places*, the list provided a thoughtful, data-driven glimpse of places where pollution had severe everyday effects—effects that could be mitigated, if anyone bothered.

PR-wise, this was a stroke of genius. People love top-ten lists. Top-five lists, top-one-hundred lists, lists of any length. Anyone who craves hits for a website need only publish an article with a headline like "Seven Most Egregiously Philandering Basketball Players," and watch the traffic flow. Blacksmith's list was no exception. The report was splashed across magazines and newspapers around the world.

But because the real point of any list, however long, is to know who's at the top of it, a lot of the coverage focused on Linfen, which had taken the number-one spot. The city became instantly notorious as the most polluted spot on the globe. (Not incidentally, I believe this report to be the original source for the picture of Sad Coal Man.) And this is the continuing source of the city's fame, fueling article after article about how Linfen is—or was, or may one day be again—the most polluted city in the world. It was the reason I'd first heard of Linfen, and the reason I was now there.

There was, however, a problem with the list.

It wasn't Blacksmith's fault, really. The report's authors clearly understood that coming up with a list of the ten most polluted places in the world was, at some level, silly. It was the same silliness I encountered when I set myself the task of choosing destinations for this book. By what standard do you make the judgment? Health effects? Contribution to climate change? Simple grossness? Blacksmith's focus—namely, industrial pollution with large affected populations in the low- and middle-income world—was tidier than mine. But even within that niche, it is ultimately fruitless to declare that the radiation in the Exclusion Zone is better or worse than the smog in Linfen. It's comparing cesium apples to carbon oranges.

To account for this, Blacksmith did something very reasonable: it refused to rank the places on the list. The report even says so, on page 6: "It was not realistic to put [the locations] into a final order from one to ten."

Instead, the list was ordered by country. *Alphabetically*.

Nobody noticed. Such distinctions are no match for a reader's desire to know *who's number one*. And so Linfen took the crown . . . because the *C* in *China* comes near the top of the Roman alphabet.

Also unnoticed was that Blacksmith intended Linfen merely as an example of its kind. "Linfen acts in the Top Ten as an example of highly polluted cities in China," reads a note on page 14. "In terms of air quality, the World Bank has been quoted as estimating that 16 of the 20 most polluted cities in the world were in China."

Have a little sympathy, then, for the citizens of Linfen, who instead of a one-in-ten or a one-in-sixteen ranking had to carry the gold medal all on their own. Meanwhile, the Russian city of Norilsk, also on the list, dodged a cannonball of bad press, simply because *R* comes after *C*.

Linfen may well have improved since those days. Blacksmith is mum on the topic of late. After a couple of years they realized that providing fodder for sensationalist headlines—and alienating local governments and industry—was not in their strategic interest. They moved on to list toxic *problems* instead of toxic places. But the taxi driver was right. By the time the 2007 list was released, Linfen was no longer number one. It had lost out to some place in Azerbaijan.

\\\\\\\\\\\

Coal pervades Linfen. It feeds the furnaces of power plants and of single-family homes. In the form of coke, it fuels the sprawling steel plant just east of downtown, a coal-fired fantasia of industrial power that is the last thing an American expects to see in the middle of a residential area. Our very casual attempts to stroll into this steelmaking city within a city were shut down right at the gate, but the guards were friendly enough to let Cecily use the bathroom just beyond the checkpoint. (I recommend trying the coffee shop's restroom first. Cecily described the one at the guard post as "horrible.")

We wandered the plant's margins, through a crowded neighborhood, poorer than the ones we had found near the drum tower. A small pack of boys bearing plastic firearms became our escort.

"Where is he from?" they asked.

"America," Cecily said.

"How long have you been traveling?"

"Three years," she answered.

Perhaps more than for the actual coal, Shanxi Province is famous for the coal bosses, a class of nouveaux riches that became astronomically wealthy as the Chinese economy took off. They were legendary for their appetites, for showing up in Beijing and buying one of everything. The most expensive watch, the most expensive car—it was all fodder for a coal boss's rapacious lifestyle. I had heard the tale, probably apocryphal, of a coal boss who, liking the looks of an apartment building under construction in Beijing, had decided to buy every unit with a southern exposure. Cecily told me that her friends would joke about marrying coal bosses, in much the same way, it seemed, that I had heard young American women joke about finding an investment banker or a hedge fund manager.

I suspect the coal bosses personified certain anxieties about the way capitalism was driving China's transformation. They were a farcical over-statement of the consumerism that was spreading through the middle class in general. And worse, the coal bosses' wealth was exploitative, in that it came from a dangerous and often illegal industry. China's coal mines were notorious for collapses and explosions, with a cost in lives that outstripped any other nation's mines.

But the golden age of the Shanxi coal boss was drawing to a close. The government had consolidated or closed thousands of coal mines, in a bid to increase efficiency and safety. And the future of the industry lay in less-developed provinces like Inner Mongolia, where huge reserves of coal waited in the ground.

Linfen isn't really a coal baron town; I hear they prefer the provincial capital of Taiyuan. But even in Linfen, a crust of luxury is overlain on the economy. There was, for example, the Audi dealership—a striking mesh-clad box that housed a sleek, museum-like showroom.

"Our customers are mostly from coal mines," said a young salesman called Yanlin. And he didn't mean the miners themselves. Industrialists liked Audis, he told us. Executives from coal mines, metal mines, coke factories. They came here to buy their cars.

A brand like Mercedes-Benz attracts too much attention, Yanlin said. Audi is a good car, very good quality, but not as gaudy. It shows they are the boss, but is a little more low-profile.

Even so, an Audi could go for two million yuan—three hundred thousand dollars. And sales were still good, even with the recent consolidation of the coal industry.

Yanlin seemed to be getting a little nervous at all the questions, so we thanked him and went to roam the showroom floor. I was less drawn to the cars themselves than to the display cases of Audi-branded accessories: leather wallets and portfolios, pens, an iPod case or two, all stamped with the quadruple circle of Audi. The placard for a handbag read, in Chinese, "This purse is a miracle."

A pair of cufflinks caught my eye. They were engraved with the logo for the Audi R8, a high-performance sports car. The face of each cufflink was mounted with a small, lacquered checkerboard of carbon fiber. This was probably a reference to carbon-fiber components used in the cars, but here, in coal country, the cufflinks took on special meaning.

"Everyone has to have their own style," said Cecily, reading the placard. "These cufflinks show your spirit and taste. Made with real carbon and stainless steel."

Were there Shanxi coal men driving around wearing cufflinks made of carbon? It was too good to be true, but one of the salespeople assured us that it was. He also told us that there were health benefits to wearing the cufflinks—the carbon in them absorbed toxins. But this harebrained theory was less interesting to me than the idea that the cufflinks were some kind of badge of honor, a Masonic ring for that brotherhood of men who are helping us seal the deal on climate change. (Order your own from the Audi Web site for $169.)

The dealership's customer service director, a young man called Jun, had taken an interest in us. He had nobody to eat lunch with that day and offered to take us out. I noticed that he drove a Nissan.

I don't make enough to buy an Audi yet, he said.

We had lunch at the Taotang Native Association, an ornate wonderland of executive schmoozing. It was a recent building, set down on a stretch of land not far from the Yao Temple and the Hua Gate, near the construction site of a huge shopping mall. We made our way through a warren of court-yards and corridors, into a small ballroom with a stage, and finally to a private room with a large circular table outfitted with a lazy Susan.

Jun ordered lavishly, without looking at the menu, and soon there were something like fifteen dishes on the table, including foie gras, shredded rabbit with cabbage, tofu, fried buns, garlic broccoli, and something Cecily translated as "specialized noodles."

Jun was twenty-eight, with an attentive face and a crown of wiry hair bursting off his head. He smoked between his measured assaults on the food, and atomized the conversation into small sections divided by the rings of his two cellphones—one white, one black—sometimes stepping out of the room to talk. The white phone was for regular calls, he said. The black one was for his most important customers. The black one he answered twenty-four hours a day.

The guy was unstoppable. "Car sales depend on personal relationships," he said, and pushed the turntable so the pile of foie gras was in front of me.

He told us that, in sales, you have to put yourself in the customers' shoes. Anticipate their needs. Become their friend. Then, when they have to choose a car, they will come to you. "Competition is very fierce here," he said. "You win not by price, but by personal relationships."

He had majored in car repair in college, but had been doing this job for four years now.

"So you went from servicing the cars to servicing the customers," I joked.

He nodded. "That's correct."

There was nothing Jun wouldn't do for his clients, whether it was helping them with personal business, running errands, or doing other favors. For one customer, he had recommended a stock pick—and promised to reimburse him if he took a loss on the investment. And he told of having to talk one powerful client down from a drunken rage after a relative received some minor injuries blamed on a faulty Audi air bag. Many of Jun's clients were rude, he said. But no matter how rude they got, he couldn't allow himself to lose his temper. In the end, the rudest ones ended up trusting him the most, because he tolerated their behavior more than anyone else would.

The most important part of his job, though, was to smoke and eat and drink without cease. He drank and smoked much more than before. Nothing got done in Shanxi without drinking, he said. It was a hard job. His family didn't like it.

The conversation had somehow turned, and suddenly, instead of telling us about his ability to manage the personalities of truculent coal men, Jun was talking about how depressed he was.

He had taken a test on the Internet, he told us. A score above fifty meant you were depressed. He scored eighty.

He was under a lot of pressure, he said. He wanted to keep making his salary. He was making five times what he had hoped for when he graduated, and had bought a house for his parents, but he was too busy to enjoy life. He couldn't relax. He had no time to socialize, and was starting to lose friends. They assumed he was avoiding them, that he thought he was better than them, because of the money he made. But that wasn't it. He was just busy. The only people he really talked to, he said, were friends he had made online, whom he would never meet in person.

He wanted to have a plan for the future, he said. He wanted to get married. But he didn't have time to think about himself, to think about his "self-strategy." The only time he could think about such things was late at night. Or while sleeping.

He lit another cigarette. All he knew was that, if he left the car business, the connections he had built in Linfen's industrial community would be valuable, whatever he did. Relationships weren't important just for selling Audis.

I wanted an audience. I wanted to smoke and eat and drink with one of the coal princes of Shanxi. I wanted Jun to help me get a glimpse of the top of the food chain.

He shook his head, a faint smile on his face. It's impossible to meet those people, he said. Then he got the check, and we rose from the table to go. But before we left, I took another bite of shredded rabbit. You have to eat the year, or it will eat you first.

\\\\\\\\\\

The closest I got to meeting a coal boss was Liu, the driver who had picked us up at the airport in Yuncheng and driven us to Linfen, and whom we had hired again several times. A middle-aged man with a sleepy expression and a wry smile, Liu and his family had started a coal mine of their own. Only in Shanxi, perhaps, does a taxi driver start a coal mine on the side. Before the recent spate of shutdowns and consolidations, the province had been riddled with small, illegal coal mines.

We had twenty workers, Liu said. Everything was done manually. But it was shut down by the government. We didn't really make any money at it.

We didn't press him. I don't know what cover story Cecily had offered— lost mental patient, maybe?—but it was likely that the Liu family mine had been illegal, and Cecily thought he would spook easily.

I liked Liu. He was considerate—and funny, although Cecily would never translate his jokes—and if he was a little cagey about his failed career as a coal boss, he was still game for adventure. Today he was taking us into the mountains to gate-crash a coal mine.

We headed west out of Linfen. It was what seemed like a sunny day, with occasional patterns on the ground that looked faintly like shadows. We could see farther down the streets than before, and the sky directly overhead was almost blue.

"Look," I said to Cecily. "Blue sky."

She looked.

"It's not blue," she said. "It's gray."

I looked again. I was pretty sure it was blue. Compared with the dingy taupe of the horizon, it was distinctly bluish.

Cecily shook her head. "I know you. You just like polluted city."

We had crossed the river and were now passing through small squares of farmland on the outskirts of town. A curtain of smog opened, and a smokestack painted with blue and white stripes loomed over us. It was Linfen Thermo Electron, a huge coal-fired power plant.

Its sudden appearance out of the haze was appropriate to the rate at which coal-fired power plants are being built in China. Depending on whom you ask, China has added them to its grid at the rate of one a week, or one every four days, or one every ten days. As we passed the plant's gate, such numbers took on a mind-boggling significance. Thermo Electron was sleek and massive, a raised fortress with soaring walls of blue metal that would have shone in the sun, had the sun chosen to shine. Yet it was only one of the countless plants that were being plopped down one after the next across the country. Enough to power China. Enough to make up for any fossil fuel you and I haven't burned.

With so many new coal-fired plants, the Chinese government was having trouble keeping the industry in balance. Record-breaking demand created spikes in the price of coal—but the Chinese government was reluctant to let power companies pass the cost increases along to their customers. So the producers simply chose to produce less power, even as coal extraction rose to record levels. That spring would see some of the worst electricity shortages in years.

Just beyond the power plant, the plain erupted into walls of bare, craggy rock. A temple had embedded itself high on a shattered rock face, a ramshackle fortress trailing a staircase down to earth.

Into the mountains. We climbed past villages. Houses had been carved out of mountain faces, rock alcoves faced with brick walls that allowed a single door and window. Piles of coal sat out front. Black smoke trickled from horizontal chimney pipes. Coal trucks rumbled forward and past, and sat in front of houses, and in repair shops. I thought of the logging trucks I had seen rumbling along BR-163 near Santarém, and about the giant sand haulers in Alberta, and of Nelson's little dump truck in Beaumont, Texas, and for a moment it seemed likely the world was composed mainly of trucks.

\\\\\\\\\\

We infiltrated by walking in the gate. Liu had sniffed out a coal mine for us. Actually, there may not have been a gate, just a narrow road leading into a broad loading pit. The loading area was a small landscape draped with a layer of coal powder an inch or two thick. A short mountain of coal sat by the battered housing of a conveyor-sorter, waiting for the afternoon's convoy of trucks to carry it away.

The experience of leaving soft footprints in a blanket of coal powder is dizzyingly similar to walking through a fresh, dry snowfall. Wavelets of black dust scatter from your feet. *It's just like snow but black*, you think—and somehow this feels profound.

A man coasted down the hill on his motorcycle, heading toward the town we had come through on our way up. Would he sound the alarm? After our warmish reception in Guiyu, it seemed that the world owed us some unfriendliness, and Cecily and I were ready to be screamed at and kicked out. But the man on the motorcycle barely gave us a look. So far so good.

We walked uphill to a set of buildings and railroad tracks that surrounded the mine's extraction mouth. The miners themselves entered through another tunnel, farther down the mountain, but this was where the coal came out, in old-fashioned mining carts similar to those you may have seen carrying Indiana Jones.

It was here, at last, that I came face to face with Sad Coal Man.

He was taking a nap. Or smoking a cigarette. Two of him were chatting with each other. There were eight or nine of him altogether. And just like in his photograph, each of him was wearing clothes darkened with coal, and had a face dusted and smeared with fine, black grit.

But though they looked just like the guy in the picture, there was something different about these sad coal men. They weren't all that sad. At worst, they seemed kind of . . . bored? They were between shifts when we walked into the work area, I think. Maybe they were waiting for something to be fixed down below. So instead of working, the not-so-sad coal men were lounging and chatting, resting in the sun, and playing with a visiting toddler. That was incongruous, I thought, a child toddling among the coal carts. Some of the nearby buildings, we learned, were housing for the aboveground workers. One of their wives had brought the toddler for a workplace visit, commuting from about thirty feet away.

Our presence had yet to raise an eyebrow, which I found disorienting. I had gotten used to being noticed. And although the attention that comes with sticking out in a foreign country makes me uncomfortable, I had lived with it for long enough now that this absence of discomfort felt pretty awkward itself.

We leaned against a girder and observed the spectacular lack of activity. Nothing came out of the mine. Nothing happened. Behind us, a woman tossed a shovelful of coal into a fire under a pot of boiling water.

"This is a state-run coal mine," Cecily said.

"In the United States there's a stereotype that government jobs are very stable," I said. "Very easy."

She nodded. "Same thing. It's why people work for the government. My parents always wanted me to work as a civil servant. We call it the *iron bowl*. Because you'll never break it." She shook her head. "Boring!"

A supervisor in a trim blue blazer wandered over to us and offered me a cigarette. The offering of cigarettes was a ritual that almost took the place of shaking hands around these parts, and I had bought a pack of my own in order to participate. But because my reflexes had yet to develop, I was always too slow on the draw. I had tried to ramp up my smoking skills before I got to China, as a sort of lung-destroying backup plan, but hadn't had the discipline. By the time I fumbled the pack out of my coat and shook it in the supervisor's face, he had already lit a cigarette of his own and offered me another in that perfect way, three filters artfully peeking out of the pack.

The supervisor happened to be from Henan Province, where Cecily had grown up. That was our in. Like any large country that hasn't had its native people replaced in the past five hundred years, China is not actually a country but a collection of subcountries, and this allows for the on-the-fly formation of much stronger alliances than come about when two people discover they're both from Cleveland. Cecily told me her Henan dialect was rusty, but clearly it was still good enough to ingratiate us with the supervisor, who gamely let us hang around his work site, instead of dragging us down the hill by our ears, as he should have.

There were limits, though. He would not let us go underground. It wasn't safe, he said, and it wasn't up to him, and the people who it *was* up to wouldn't let us go below either. Besides, he said, the mine was no place for a woman. I didn't have to look to know that Cecily was making a face.

A pair of miners wandered over to refuse my cigarettes. The supervisor gestured at me and asked Cecily a question.

"Let me guess," I said. "He's asking if I'm married."

"Yeah," Cecily said.

I grabbed my head. "Really?!"

"No, no," she said. "I'm joking with you." What he actually wanted to know was why we were there.

This time, I thought, we should deploy *high school teacher*, a new role that I thought was actually plausible. But Cecily went ahead with something like *professor of mining engineering at a major American university*.

"What if they start asking me questions?" I whispered.

Don't worry, Cecily said. We just have to tell them something.

In the gear house behind us, a large wheel fitted with a cable began to turn. The two lengths of the cable descended into the tunnel-like mouth of the mine; each end was connected to a train of five mining carts. If they ever got started, one set of carts would descend, empty, while the other rose up from belowground, full of coal. But that was some time off. There was no coal to be seen, except for everywhere. The wheel was only turning so the crew could apportion the slack on the lines.

The track on which we stood ran through nearly five hundred feet of tunnel to reach the loading area, which the supervisor told us was located three hundred vertical feet belowground. Several hundred miners were down there, powering their way through seams of coal. Workers at the mine usually got a day off for every ten they worked. But lately, he said, they had been working constantly, week after week without any days off.

The mine produced a thousand tons a day, which was on the small side, now that so many illegal operations had been shut down. Mines with a daily production of three or four thousand tons a day were not uncommon. In a few years, the supervisor told us, this mine would either be retrofitted with new systems to replace its antiquated technology or be shut down entirely.

In this, it was part of a grand shift in China's coal industry. New mines were being built with more recent technology; mines like this one, which had been open since the 1940s, were being phased out. A similar shift is happening across a wide range of Chinese industries, a shift that promises to restructure the country's economic and industrial landscape. Heavy industry is moving inland, to areas where both natural resources and cheap labor are plentiful. Just as the United States and Europe have sent much of their heavy industry overseas, economies within China are bifurcating between industrialized and developing, between high and low income. The tidal wave of industry is moving farther and farther inland from the coast. In its wake stand places like Shanghai and Shenzhen, which have been transformed into resource-hungry approximations of the countries whose consumerist economies they have emulated.

"Do you use this kind of technology in America?"

It was the supervisor. He nodded toward the mining carts, the cables, the giant wheel that drove it all.

I had by now perfected the gesture: a faltering, circular motion with my head that was neither a shake nor a nod.

"Somewhat," I said.

To kill time while things got going, Cecily and I walked up the road that led over the waste rock pile, a mountain of shattered scree that partially filled the narrow neck of the valley. There, accompanied by two men with strong Shanxi accents, we looked over the vista of the mining operation: the dorms to the right, the throat of the mine tunnel to the left, and beyond it the hopper and conveyor system, which separated the coal from the waste rock.

One of the two, a rosy-cheeked man in his thirties, told us he was the manager of the explosives storehouse. He had previously worked underground as a miner, a job that had paid more. Cecily asked about his wages, and told me they compared favorably to the salary a recent college graduate could hope for in Beijing.

I found it suggestive about mine safety that someone would prefer to make less money, and to work with explosives, than to be underground.

"It's safer than it was," the man said. "But it's still not that safe."

Our other companion was an elderly man who said he had worked in the mine for several decades. Now he was retired.

"Before the 1980s," he said, "we did everything manually. We dug with tools. It was hard work. It's much better now. It's all machines."

"Did you like the work?" I asked. "Was it good work?"

"It was *hard* work," he said. "It doesn't matter whether I liked it."

Whether or not he had liked the work, though, he had decided to spend his retirement here. His children worked in the mine, and he saw no reason to leave. He smiled as he looked down over the mining complex, and not for the first time, I reflected that there was more than one kind of health.

The explosives man broke the spell. "Why did you come here?" he said, laughing. "It's so dirty and black!"

Back at the extraction area, below, I stood by the cave-like mouth of the tunnel for a long time—hours, it seemed—and smoked, and lay on a pile of corrugated sheeting in the sun, and smoked, and eventually wondered if anything was ever going to happen.

The tunnel mouth had walls and a metal ceiling, and a sign over it with two golden characters, and the tracks in the ground, and it all sloped down at a steep grade away from the daylight, to a second, deeper, older mouth, a tunnel mouth built out of stone, and then disappeared into the earth. It was a tunnel, but after staring at it for an hour, I decided that really it was a cave in the end.

I tried to offer a cigarette to the guy next to me, and was again too slow, and again accepted one of his.

Then there was a grinding sound, not quite a rumble, and the cable went tight, and the man whose cigarette I was smoking went and stood on the tracks. The grinding grew louder and louder, and everyone started paying somewhat more attention.

At last the tunnel roared, and a train of five mining carts shot out of its mouth. I felt the urge to dive for safety, even though I wasn't in the way. The man standing on the tracks—who *was* in the way—floated onto the leading cart, as casually as if he were stepping onto a streetcar. He reached down with lackadaisical precision and removed a thick metal pin to release the tow cable. The carts moved fast, spiriting him backward away from the tunnel mouth, and as they went by, I saw it. The coal. It filled each cart to the rim, in shining, sticky, dribbling mounds.

The cable-release man hopped off and another man stepped on. He threw aside another pin and the link that connected the lead cart to the train. The cast-off gear hit the black ground with a deep, metal clank. Another worker had put a block on the track to stop the following carts, and now the lead man coasted free, riding his cart along a wide arc of track, all the way to the hopper, where yet another man waited, also with coal on his face, also wearing a jacket and sweater, work gloves and trousers, dark with coal—everything black with the stuff. The lead man stepped off his cart just as it crashed into the hopper, and the two men pushed on the levers, using the cart's momentum to upend it. They dumped its contents into a chute, where the rocks were shaken out of it, and then the coal was sent along the belt of the conveyor-sorter and onto the heap, where a loader was now scooping huge shovelfuls into a truck.

The cart crashed upright, and the workers pulled it backward, steep into the task, grimacing it into motion until the lead man could pull it back toward the tunnel on his own, and onto a side track where the cart would await its next descent into the mine. Already the second cart had been detached, and shouldered down the faint slope by the next member of the crew, the worker who had been napping when we arrived, his eyes still drowsy now, as he pushed his ton toward the hopper.

We had hung around all afternoon, and watched the workers crack jokes, and shared the international ritual of looking at pictures on the display of a digital camera, and now we saw the workers work. Coal kept shooting out of the mine, in five-cart loads, and the not-so-sad coal men swung through their practiced motions. They would have made a great advertisement for Chinese Communism in its pre-capitalist days. Each worker had his role, each role its place in the chain, a choreography of labor, skilled in a way that only unskilled work can be. They brimmed with rugged, coal-stained intelligence, pausing between mine trains to smoke and to talk. Sad Coal Man was not debased and morose. He was sharp. He was witty. He smiled in the sun when he had a few minutes to himself. Maybe he was just glad to be aboveground.

\\\\\\\\\\

In the lobby of the hotel, Cecily tried the automatic shoe polisher on her sneakers, completely black from our afternoon at the mine. It had no effect except to blacken the spinning brush of the machine.

I went to my room. I took off my clothes. They were covered in black stains, although I had pushed no carts, handled no coal. It just happens when you're at a coal mine. Liu had driven down to town ahead of us, so at lunch-time the miners gave us rides down the mountain on their motorcycles. And as soon as you rub shoulders with a coal miner—although you may look clean by comparison—you will find black dust on your every surface, in your pores, under your fingernails.

I washed my face, and stared at the smear of black on the washcloth, and sat on the edge of the bed, and I missed the Doctor. I missed her.

I thought of the motorcycle ride down the valley. We had run with the engine off, because even miners like to save gas, coasting down the steep mountainside at speed, the wind pulling tears out of my eyes, and when I got off the bike, the front of my jacket was streaked black where I had leaned against Sad Coal Man as he drove.

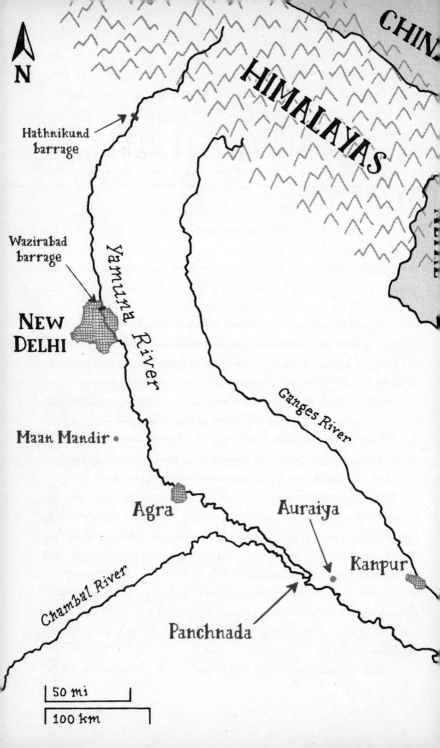

SEVEN

THE GODS OF SEWAGE

After Sati killed herself, her husband was inconsolable. It was Sati who had convinced him to love, and who had taught him desire. It was for Sati that he had emerged from a life of austerity and isolation to be part of the world. And it was for his honor that she had thrown herself on the pyre.

He pulled her body from the fire and carried it for days, wandering, crazed with grief and rage. Because he was Shiva, and a god, his fury was destruction—a chaos that threatened to engulf the whole world. Vishnu went to calm him down, dismembering Sati's corpse as Shiva carried it. (Gods have their ways.) People still worship at the places where Sati's body parts fell.

Empty-handed, Shiva went to the river, to the Yamuna. Yamuna, daughter of the sun, twin sister of death, goddess of love and compassion. He bathed in the river, and as the madness of his grief cooled, it scorched the water black.

This, they say, explains the color of the Yamuna—so distinct from the milky waters of the Ganges. The holy Yamuna is a river that accepts and dilutes grief and rage, a fount of love and understanding for everyone from the gods on down. Maybe it is mythologically appropriate, then, that it accepts so much else.

India is full of holy rivers, and even derives its name from a river. It is the *land beyond the Indus*, a river whose own name, just to be safe, derives from an ancient Sanskrit word for river. And as with Hindu deities, so with Indian waterways. The name of the game is multiplicity. Each is the incarnation or avatar or consort or child of every other, and there is hardly a creek in the subcontinent that can escape the burden of some pretty hard-core metaphysical freight. How holy are India's rivers? So holy that even certain bodies of water in Queens are also holy. So holy that you can't spill your drink without worrying that someone will show up to venerate it.

The Ganges—or *Ganga*, as it is called in India—is, by many accounts, the holiest of all. Heart of Varanasi, consort of Vishnu, flowing through the hair of Shiva, etc., etc. It is the apotheosis and parent of all other rivers. And it was on the Ganga's banks, nearly a decade earlier, that I had first seen the light as a pollution tourist. I had lived in New Delhi for six months and had happened to visit Kanpur, where the Ganga received a crippling infusion of industrial effluent and municipal sewage. It was supposedly the most polluted city in India. But I *liked* visiting Kanpur. I liked how you could walk from the tanneries to the river, from the open sewers to the farms, and see for yourself how they were all connected. I liked how you could stand on the banks of the reeking Ganga, almost as sludgy as it was holy, and watch pilgrims take their holy baths, confident in the purifying power of the impure water. All this, and cheap hotels. Yet in the guidebooks, Kanpur didn't exist.

Well *that's* not fair, I'd thought.

And in Delhi, I had met a different species of environmentalism from that in the United States. Back home, however much you thought you cared about the environment, it was an impersonal concern. After all, your daily surroundings, whether in suburb or city, were likely to be pleasant, or at least clean, or at least nontoxic. In India, though, environmentalism was more than an abstract moral value. It was more than a way to signal your politics and your socioeconomic status. Here, in the daily confrontation with poor air and adulterated drinking water, it took on the urgency of a civil rights struggle. Only in the polluted places could you properly understand what was at stake.

This time I skipped Kanpur. Skipped Ganga. It might be India's holiest river, but the Yamuna is its most polluted, and I had priorities. I wanted to know why, with all the Hindu rumpus about rivers, a river goddess can't actually catch a break. For although the Yamuna might be a goddess, by the time she leaves Delhi, she is no longer a river.

\\\\\\\\\\\\

I hadn't gone home. I had none. I had come straight from China. From Lin-fen to Beijing, from Beijing to Shanghai, from Shanghai to Delhi. Delhi, where, not five minutes from the airport, the cabdriver resumed where the Han family had left off.

"You are married?" he asked.

Had entire continents been populated only to make me say it? I was alone. Not with the Doctor, not newly married, but alone, and alone, and alone.

"No," I said. "Are you?"

He nodded. He had a child, too.

"Your country, all love marriages. No arranged marriages. This is good," he said. "Arranged marriage, father and mother choose the girl. You choose different girl."

"You had an arranged marriage?" I asked.

"Yes."

"Do you love your wife?"

"Yes," he said. His eyes were on the road. "But I loved another woman."

\\\\\\\\\\\\

India. Land of contrasts.

That's what you're not supposed to write about India. But nobody can help it. Even the most sophisticated people will write thoughtful, evocative prose that still amounts to *India, land of contrasts*. What are they trying to say? Is there no contrast anywhere else in the world? I think what they mean is *India, less tidy and homogenous than I'm used to.*

I had loved Delhi when I'd lived here. Loved the noise, the smells, the energy of the street. That's what I wrote home about. I had reveled even in the simple tumult of buying a train ticket. But as the truism has it, a traveler's writings say more about the traveler than about the place traveled to. Before, I had found dusty blossoms of curiosity and independence on every corner. Now, years later, I saw Delhi again and wondered if I could just sleep through it. Through drab, mediocre Delhi.

But it wasn't just me. Delhi had changed in the past decade. At least, that's what people told me.

"Oh!" they would say. "Delhi has changed so much!" Even the autorickshaw drivers, if they spoke English, would tell me how bad the traffic had become, as if there were no traffic jams in Delhi in 2002. And those same autorickshaw drivers still pouted when you tried not to let them rip you off as fully as they wanted.

So Delhi was still recognizably Delhi. But it was true—there had been some restyling. Its elite shopping malls more convincingly suggested that you might be in America. The Evergreen Sweet House restaurant now had three floors, and air-conditioning. The city's upscale neighborhoods were marginally tidier than before, and disappointingly free of wildlife. Street animals used to be half the fun in Delhi, but now you've got to work to bring your clichés to life, and you're down by Tughlaqabad before you can find a pair of cows blocking the road.

The most obvious change was the Delhi Metro, whose routes had burrowed through the city far more rapidly and effectively than anyone could have expected. It now ran all the way down to the satellite city of Gurgaon, about ten miles to the southeast. A subway to Gurgaon, imagine! The success of the Metro seemed to have taken the city by surprise. In a land where public works are so often lumbering, ineffectual, and corrupt, the subway was clean, efficient, and cheap.

As for the Yamuna, I had no idea if it had changed. Its banks lay only a few miles from where I had lived, but at the time I had been only dimly aware that a river even existed in Delhi. It was an appropriate ignorance, though. Delhi had long since turned its back on the Yamuna. Now the river played a part in the city's life only as an object of neglect and disgust.

\\\\\\\\\\\

On the riverbank, I gave a man called Ravinder a few hundred rupees and we went out in his flat-bottomed wooden rowboat. Sitting next to Kakoli, my translator for the day, I peered over the edge as Ravinder worked the oars. The surface of the water was a dark gradient of billowing grays interrupted by little bubbles. Methane, I assumed. We coasted into a stretch of water spread with an unfamiliar film, not quite as colorful as a petroleum rainbow, not quite as thick as the skin of milk on a boiling pot of chai. Lumpy black gobbets dotted its surface. We needed only our noses to understand that the water was dark with more than Shiva's grief. We were floating not on a river, but on a great urban outflow, a stream of human sewage that was standing in for the river that had dug the channel.

The Yamuna was full of shit.

It gets this way in stages. Emerging clean from the Himalayas, the river receives periodic doses of sewage and industrial runoff as it crosses the plain. Then, about 140 miles upstream from Delhi, it meets the Hathnikund Barrage—a multi-gated dam built to control the river's flow. At Hathnikund, the greater part of the river's water is diverted into the Eastern and Western Yamuna Canals. These canals, both hundreds of years old, were originally devised for irrigation, but an increasing amount of the water they divert is used for city water supplies, especially Delhi's. The city's population has grown more than 600 percent over the past fifty years, drastically increasing its water use.

In India, as in so many places, tension over water is the driving force behind an incredible swath of environmental and political problems. In this case, to make up for water spirited away by the megalopolis downstream, farmers in the region pump massive volumes of groundwater. The overextraction is so intense that it has lowered the water table to below the level of the riverbed itself, meaning that south of Hathnikund, the Yamuna simply percolates straight into the ground. The river runs dry. Except during the several months of the monsoon, the Yamuna essentially ceases to exist as it approaches Delhi.

Because it would otherwise disappear into the riverbed, water extracted for Delhi is transported via the Munak Escape, a fork of the Western Yamuna Canal that itself receives a helping of industrial waste and domestic sewage. The water then collects behind the Wazirabad Barrage, on Delhi's northern margin. (Here it is augmented with water brought from the Ganga, of all places, making the Ganga a tributary of one of its own tributaries.)

Thanks to these inputs, there is water in the river at Wazirabad. But this water does not flow south into Delhi, as the river once did. Instead, it is pumped out and treated, becoming the basis of the city's water supply.

Nevertheless, there is water downstream of the Wazirabad Barrage, flowing the fourteen miles through the heart of Delhi. For this stretch, the Yamuna takes the city itself as its source, receiving something close to a billion gallons of wastewater each day, the vast majority of it domestic sewage, and more than half of it completely untreated.

So when local activists refer to the Yamuna as a *sewage canal*, as they do, it is no figure of speech. Except during the monsoon, there would be no river in Delhi without this wastewater.

Nor is it much of an exaggeration when people refer to the Yamuna as *dead*. The river's level of dissolved oxygen (a good indicator of its capacity to sustain life) here falls to approximately one-tenth of the minimum government standard. Coliform levels (which indicate a waterway's microbial danger) are incredibly high. The Indian government's upper limit for safe bathing is five hundred coliforms per hundred milliliters of water. At points in Delhi, though, the coliform count has exceeded seventeen *million*.

The oarlocks squeaked and knocked as Ravinder worked the oars. He wore a Levi Strauss T-shirt and blue track pants. The lifeless river was placid, almost pleasant. A light breeze took the edge off the sewage smell.

"Who told you this is water?" he said. He told us that when he was young, he had been able to see to the bottom of the river. Now, though, you could barely see a foot deep, and clouds of inky muck eddied against the surface as we passed through shallow areas, the ends of the oars black where they had touched the bottom.

Ravinder had grown up on the banks of the Yamuna, and still lived in one of the city's few riverfront neighborhoods. And in his thirty-odd years he had seen the river change. "There were lots of tortoises, but people sold them off. There were fish, and snakes," he said. "But now it's just a drain." Although he lived mere steps away from the river, he neither bathed in it nor allowed his family to. Only in July and August, during the annual floods of the monsoon, would they get in the water. "During that period, the river becomes very beautiful," he said. "But within a month, it's over."

Ravinder earned his money by taking people out to the center of the river to drop offerings or cremation ashes in the water. Sometimes he made a thousand rupees in a day—about twenty dollars. Sometimes he made nothing.

"I took two people out on the river earlier today to drop eighty kilos of charcoal in the water," he said. "A priest told them to. They invoked the name of the sun, and of Yamuna, and dropped handfuls of charcoal into the river. Then they dumped the rest out of the bags. Tomorrow morning, I'm taking a couple to put a hundred and twenty fish into the river."

"Living fish?" I asked.

"Living fish," he said.

Kakoli shook her head. "Those fish will die."

A printed picture of a blue-skinned deity came floating downstream. Before I could make out if it was Shiva or Krishna, the oar struck it on the downstroke, folding the image and plunging it into the black water.

A pair of men were bathing on the riverbank. A gull flew over our heads. Upriver we saw a hawk, a tern. Over Ravinder's right shoulder, Nigambodh Ghat was coming into view—the cremation ground. A trio of pyres burned on the shore, braiding the air into thick tangles of heat.

The cremation ground is one of the few lively spots on the riverside, and a surprisingly relaxing place to spend the morning. Kakoli and I had visited before going downriver to find Ravinder. We had sat on a large concrete step and watched a group of young men build one of the pyres now burning. (There was a gas-fired crematorium just down the bank, but no person in his right mind wants to be cremated in a dank, indoor, gas crematorium. Not if your family can afford the wood to burn you on the riverbank.)

On a low pallet, a man lay wrapped in white cloth, his head exposed. His face was old. He was dead. The younger generation dribbled water on him from a plastic bottle and sprinkled dirt over his body. Then they finished building the pyre, leaning planks and branches against the man until they had formed a teepee of wood four or five feet tall. It was ten in the morning.

"In Calcutta, people still go to bathe in the river," Kakoli said. "Even wealthier people. But in Delhi, people will not look at it. People will only come to the river to use it as a cremation ground."

A young man in black trousers and a red sweater walked around the pyre, holding a thin strip of burning wood. It was the dead man's son, I assumed. He stopped at the head of the pyre and lit it near the ground. A thin trail of smoke trickled out. That's where we all go—not back to dust, but into the atmosphere, to join our emissions. The young man and his five companions then retired to one of the concrete tiers facing the bank and began their wait, chatting casually. It would take several hours for the pyre to burn.

Riding by the pyres in Ravinder's boat, I now noticed a pair of men standing knee-deep in the water, mucking out scoops of mud. They were collecting ashes that had been cast into the water from the riverbank. A cremated person may have been wearing rings, or been adorned with other precious objects, as they were placed on their pyre. Now these men were poring through their sodden ashes to see what they could find. Gold fillings, maybe?

I asked Ravinder if this wasn't, you know, *bad manners*.

He frowned, looking at the men on the bank. No, he said. It's not seen as disrespectful.

South of the cremation ground, we crossed wakes with a dark-skinned woman wearing an olive-colored sari. She was squatting on a large plastic bag stuffed with scraps of polystyrene foam and mounted with a square wooden frame. Her raft listed forward steeply to where she hunkered on its edge, working the water with a single, short oar.

Her name was Mamta. She lived with her husband on the opposite bank. They made their living combing the margins of the river for paper, plastic, anything they could sell to the recyclers. Her raft was littered with the morning's haul: several coconuts, a few paper booklets, and a single plastic sandal.

She stared at the water as she answered my questions. They had been in Delhi for ten or fifteen years, she said. Eight years ago, the government had pushed them out of the shantytown they had lived in. Now, they lived in a temporary shack on the floodplain. When the river rose each year with the monsoon, they had to retreat with the land.

When I asked how old she was, she hesitated. "I don't know," she said. "But I think I'm twenty-five or twenty-six." Then she continued upriver, raking her oar through the mat of flowers and trash that clung to the bank.

Ravinder sent us back out to the middle of the river, where he left off rowing and let us drift. He crossed his legs and opened a packet of tobacco. "So many people migrated to Delhi," he said. "The waste going into the river has grown and grown with the city. But Yamuna is one. It has not multiplied."

He still believed in the river, though. Yamuna was a goddess, he said. He might go for a week and a half without earning any money at all—only to make up for it in a single day's work. The Yamuna didn't take, he said. It gave.

With that, he put some tobacco in his mouth and we drifted for a while longer, spinning quiet circles in the breeze.

\\\\\\\\\\

Where there are rivers or lakes in India, there are ghats: wide riverfront stairs that lead down to the water. Ghats are an indispensable part of the sacred Hindu love affair with water, and through history they have been places for worship, and worshipful bathing, and non-worshipful swimming, and for doing the laundry, and for cremating the dead—as at Nigambodh Ghat—and for pretty much anything else you might want to do at the riverside. But Delhi has few ghats. It is a city of sixteen million with barely any places, ghat or otherwise, where people interact with the river. I went looking for any that were left.

At the south end of its Delhi segment, the Yamuna is again made to jump its channel. The Okhla Barrage shunts it into the Agra Canal, through which it is destined to become the Taj Mahal city's unenviable water supply. Just upstream of the barrage is the riverfront park of Kalindi Kunj. Unlike many riverfront parks, though, Kalindi Kunj offers no actual frontage to its river. A fence encloses the park, keeping visitors away from the actual river, which sits quiet and littered with trash. Ill-disposed to climb an eight-foot-tall fence topped with spikes, I resigned myself to wandering the leafy confines of the park.

The place was crawling with young couples in the throes of passionate hand-holding. With every corner I turned, I almost stepped on a pair of sweethearts. In a city where young couples have no apartments or cars of their own to disappear into, they go to the parks. It is so common here for couples to meet each other in parks or at historical monuments that it sometimes seems that these places have been designated by the city government as official make-out spots.

It ought to have sent me into a lovelorn tailspin, like everything else did. Instead, it was a respite. Since arriving in Delhi, I had been preoccupied with how Indian men and women interacted in public—or how they didn't. It's safe to say that the vast majority of Indians live under very conservative sexual mores, and it had been depressing the hell out of me.

Maybe it was the astounding numbers I had recently heard about child sexual abuse in India. The country is home to more than four hundred million children, nearly a fifth of the world's below-eighteen population, and according to the government more than half are sexually abused. Incredible India, land of contrasts, awash in brutality.

I thought of this every time I boarded the Delhi Metro. There are separate cars for men and women—which itself says something—and as we filed on, I would think of those children growing up, of what my fellow male passengers must be carrying inside them, and of what they must have done, and of hundreds of millions of lives distorted by such epidemic violence and rape. By the time the train left the station, I'd have convinced myself that men were born only for cruelty, and that no living person, woman or man, would ever escape our planet-eating vortex of betrayal and isolation.

I was down.

In Kalindi Kunj, though, it was different. Maybe there was hope—just a little—for loving coexistence between the human species. Every second tree hosted a couple that sat at its base, talking quietly, laughing, holding hands, kissing. Everyone was running their hands through someone's hair. Everyone was cradling the head of their beloved in their lap. If the woman wore a sari, she might drape its veil over both their heads. Who knows what went on in those micro-zones of privacy? Everyone was smitten. On a perfect spring day, thirty yards upwind from the shittiest stretch of river in the world, I believed in love for a little while.

\\\\\\\\\\

There once were ghats up by the ISBT highway bridge, but for no good reason the city government ripped them out in the early 2000s. Now the overpass itself serves as a kind of high-altitude, drive-thru ghat. As on other bridges over the river, people pause day and night to throw offerings or trash into the water. It's hard to tell the worship from the littering.

My friend Mansi brought her camera, and we spent a morning underneath the overpass, where a slope of packed dirt led down to the river. Every minute or two, an untidy rain of flowers would sift down from the bridge, or a full plastic bag would hit the water with a dank *plop*. We would look heavenward, sometimes catching a motorcycle helmet peering down from the railing. The city government had erected fences on most bridges to keep people from throwing over so many offerings; invariably the fences become tufted with flowers and bags that snag as someone tries to throw them over. Here, though, people had found an unprotected spot where they could throw their offerings unhindered. It was the same kind of unceremonious ceremony that I had seen at the cremation grounds, a sacredness that had no use for aesthetics.

And as with the cremation grounds, anything of value that goes into the water here must also come out. Wherever offerings are made, there are coin collectors, men who scour the river bottom with their hands. Although they are called coin collectors, they are comprehensive in their religious recycling, and actually collect anything that can be sold or reused.

The sun had just come up, murky over the Yamuna, and on the bank four collectors were finishing their morning chores before getting down to work.

"In the summer," one of them told me, "the smell gets so strong here, your eyes water." His name was Jagdish, and he had been in the reverse-offering business for nearly twenty years, since he was a teenager. He made enough to support his wife and ten-year-old daughter.

Jagdish reeled off a list of what you could find in the water here: gold and silver rings, gold chains hung with devotional pendants, coins with images of gods. But only once in a while was the score so good. "If that happened every day," he said, "I wouldn't be here." When he found coal, he sold it to the men who ironed clothes on the side of the road. When he found paper, he sold it for recycling. Coconuts he sold to people to sell on the street, or to be pressed for coconut oil if they were dry.

When you make an offering to the Yamuna, then, you are not making a permanent transfer of spiritual wealth, but playing part in a cycle, leaving tributes that will go into the river this morning only to be fished out, sold again, and reoffered this afternoon.

Jagdish worked this part of the riverbank with his brother and two other men, and while Jagdish lived five or six kilometers away, his brother Govind lived here by the water, in a tiny, tent-like shack. Govind, a friendly man in a green baseball cap, was also in his late thirties. He explained that because the water was too dark to see through, the collectors worked by touch, bringing handfuls of mud off the bottom to inspect. Govind wasn't a good swimmer, so he only waded in to his neck. His brother did the diving, when it was necessary.

A bag of trash or offerings dropped from the overpass. In the dirt, Jagdish's pet monkey, Rani, was lying on top of his dog, Michael. Rani idly scratched the snoozing dog's stomach, a picture of interspecies peace. This was the kind of symbiotic friendship the human race needed with the rest of the natural world, I thought. But then Rani started picking at Michael's anus, and he snarled and kicked her off.

Like the boatman Ravinder and the workers at the cremation grounds, Jagdish and Govind and their colleagues were among the last people in Delhi for whom the Yamuna was a life-giver not merely in a spiritual sense but in a practical one. And Govind told us he liked the work. "We're our own boss," he said. "We go in whenever we want. We're here tension-free."

When I asked him if he was religious, he shrugged. "Because the world follows God, we have to follow God, too," he said. I wasn't sure if that meant he was a devotee or not. Did they make offerings? He waggled his head. Sometimes they would give flowers or incense. But that was it.

"We take it out," he said. "We don't put it in."

India's credentials as a pollution superpower go beyond its rivers. There are the astounding shipbreaking beaches of Alang, and the lead smelters of Tiljala. And let's not forget Kanpur, with its tannery effluent, rich in heavy metals. All of South Asia, really, is a wonderland of untreated toxic waste. And while India's per capita carbon emissions are still low, its growing economy and the fact that there are 1.2 billion of those *capitas* mean that it is still a huge source of climate-changing gases.

The irony is that, in terms of environmental law, India is extremely advanced. Its very constitution mandates that "the State shall endeavour to protect and improve the environment and to safeguard the forests and wildlife of the country." As if that weren't enough to make an American environmentalist weak at the knees, it goes on to declare that "it shall be the duty of every citizen of India to protect and improve the natural environment, including forests, lakes, rivers and wildlife, and to have compassion for living creatures." And it's backed up by an activist supreme court that issues binding rulings on specific problems. Sounds like paradise.

Yet the results aren't great. Bharat Lal Seth, a researcher and writer at the Delhi-based Centre for Science and Environment, told me that although the court system is activist, this is merely because the executive branches of government shy away from taking action, leaving it to the judiciary to issue edicts. But rulings are useless on their own.

"The judiciary feeds the [environmental] movement, and the movement feeds the judiciary," Seth said, sitting in CSE's open-air lunchroom. "You get a landmark ruling, and . . . what's going to come of it?" The very fact that the Indian government doesn't feel threatened or bound by such decisions makes it easier for the court to issue them.

Seth had put me in touch with R. C. Trivedi, a retired engineer from the Central Pollution Control Board, who joined us in the CSE canteen. He was a small, friendly man with rectangular glasses and a short, scruffy beard, and probably knew more about the Yamuna's problems than anyone else in the country. Even after a thirty-year career, he exuded enthusiasm for the details of India's water supply and wastewater system. He smiled when he talked.

Before long, Trivedi was sketching a tangled diagram of the Yamuna in my notebook, reeling off numbers for biochemical oxygen demand and flow rate, and marking off the river's segments, from the still-flourishing Himalayan stretch, to the dry river below Hathnikund, to the Delhi segment— "basically an oxidation pond," he said—and finally the "eutrophicated" lower stretch, where the nutrients from decomposing sewage lead to algae blooms and oxygen depletion. "A lot of fish kill, we observe," he said, tapping on his newly drawn map. The eutrophicated segment runs for more than three hundred miles, until finally the Chambal, the Banas, and the Sind Rivers join it. There, he said, "it is good dilution. After that, Yamuna is quite clean."

Listening to Trivedi and Seth, I could see that the brutal irony of the Yamuna's situation was not only that its holiness did nothing to protect it, nor that India's tradition of environmental law was so out of joint with the actual state of its environment. The worst part was that, incredibly, cleaning up the country's rivers had for years been a major government priority. There was the Ganga Action Plan (or GAP, begun in 1985), and the Yamuna Action Plan (YAP, 1993), and the National River Conservation Plan (1995), and YAP II (2005), and YAP III (2011), among many other programs and plans, many of which continue to this day. Such programs had received massive funding, more than half a billion dollars over the previous two decades. Most of it had been spent on the construction of sewage treatment infrastructure.

At best, it had been a vast reenactment of the coin collectors' work, with the government pouring billions of rupees into the rivers, and builders of infrastructure standing by to dredge the money out.

The problem with this approach was that building sewage treatment plants was simply not enough. "We are spending huge amounts of money from the World Bank, from all other sources, taking loans," Trivedi said. But little of the wastewater infrastructure created with that money actually worked. "You have taken the loan and created it, and they don't have the money to operate it! It can work only when there is continuous flow of funds." He shook his head, smiling. "When you create a sewage treatment plant, you first figure out how it will work for twenty or thirty years. But we never looked at that. We just implemented the YAP."

"Which has no effect," I hazarded.

"Which has no effect," he confirmed.

Because Delhi doesn't charge for sewage treatment, there is no flow of funds to sustain the treatment plants. Not that most people in Delhi could afford sewage treatment fees in the first place. A further problem is the helter-skelter pattern of development in the city. A large proportion of Delhi's neighborhoods have sprouted up unplanned, without any thought for how services like water and sewage treatment could be delivered, even if they were affordable. Sewage treatment plants built with YAP funds were therefore placed where there was room for the plants, not where there was sewage to be treated.

Trivedi thought any viable solution had to address the depletion of groundwater in the river basin. That meant promoting rainwater harvesting, a practice with deep traditional roots in India. Village ponds and earthen bunds can allow monsoonal water to stand long enough for it to seep into the ground and recharge the depleted water table. "Thereby, we can reduce the depletion of the groundwater table in the entire catchment area," Trivedi said. "And if the water table comes up, all the rivers will start flowing again."

Having told me how to heal every river in India, he put his hands on the table. There was, I knew, another shoe to drop.

The rainwater harvesting, I asked. Was that something that would happen at the local level?

"Yes," he said. "But government always spends money on big, big projects. When people suggest something small, like five thousand dollars for a small reservoir or village pond . . ." He trailed off, still smiling. "They say, 'No, no, no. This is very small.'"

\\\\\\\\\\

The one place where Delhi retains a bit of the river life that it ought to have is Ram Ghat, which clings to the west side of the river immediately below the Wazirabad Barrage. It is behind this barrage, which doubles as a bridge to east Delhi, that the city's drinking water supply collects.

Ram Ghat is a bank of broad stairs dropping precipitously to the river from a wooded area next to the road. The upstream edge of the ghat abuts the barrage, itself several stories tall. Thick concrete pylons support its roadway, with metal doors in between, to hold back the upstream part of the river. In monsoon season, large volumes of water are allowed through, but on the day Mansi and I visited, all the doors were closed but for one, and even it was open only a crack. Several boys laughed and swam in the minor waterfall that spilled from its edge. Because we were upstream of the sewage drains that emptied into the river, the water here was brighter and clearer, and free of those unmentionable floating clumps. On the far side of the river we could see modest fields of vegetables. There were small fields like this up and down the floodplain, even in Delhi.

At the top of the stairs, a man wearing office clothes bought a tiny tray of birdseed from a vendor, placed it in front of some ravens on the parapet, and prayed. On the submerged steps at the bottom, a boy lingered knee-deep in the river, collecting plastic bags and scraps of wood. A few yards downriver, a woman heaved a two-foot-tall idol of Ganesh into the water. By the time his elephant-headed form had disappeared under the surface, she was already starting the climb back, dusting off her hands as she went.

I walked down the tall stairs to the water. On the bottom step, a man in a white undershirt was dragging a magnet through the water. The coin collectors were innovating. "To live, you have to do something," he said, in the universal wisdom offered to journalists who ask people about their humble, dangerous, or generally crummy jobs. And there were worse ways to spend your life than wandering up and down Ram Ghat in your shorts.

On the lowest step, I hunkered by the water. I wasn't about to take a holy dip, as they call it, but this seemed like the cleanest spot on Delhi's riverbank to get tight with the goddess of love. If it was good enough for Shiva, it was good enough for my tiny, writhing knot of a heart.

I put a hand in the water. Minute forms darted away. Water bugs. Something still lived in the Yamuna. Under the heat of the day, the river was cool against my skin. Coliform-rich, but refreshing. I lifted a handful of water. How much of this was Ganga? How much from the Munak Escape? How much had diffused its way upstream from the nearest sewage outflow? I poured it over my head. Yamuna's all-encompassing love dribbled through my hair, down the back of my neck, and soaked into the collar of my shirt.

A woman with a big white sack landed heavily on the lower step. Her daughter was with her. Together they upended the sack. Flowers and small pots tumbled out, along with what looked like disposable food trays: the leavings of some devotion performed elsewhere, which would only be completed once they had drowned the ritual scraps. Another couple overturned a bag of charcoal. Hydrocarbon rainbows spread across the water. A pair of boys standing in the river immediately started picking out the hunks.

But coins and charcoal were not the only things that got fished out at Ram Ghat. At the top of the stairs, we met Abdul Sattar, sitting cross-legged on a small rug he had rolled out on a shady bit of the parapet. He was in his mid-forties, and wore a black sweatshirt and a pencil mustache.

Sattar was the self-appointed lifeguard of Ram Ghat. By vocation he was a boatman, like Ravinder, but that was auxiliary to his real passion, which was pulling attempted suicides out of the river. He had been doing it for more than twenty-five years.

With Mansi translating, I asked him if a lot of people tried to kill themselves there. He waved his head emphatically. "Bahot," he said. *A lot.* We were only a week and a half into March, and there had already been two attempts this month.

"I didn't let it happen," Sattar said. "I can see them coming in. They generally look distressed." He had a crew of youngsters who hung out by the river. Whenever he spotted someone who looked upset, he would direct his helpers to follow the person around, so a rescuer would be close at hand in the case of a suicide attempt.

Sattar provided his services for free. And why not? All he had to do was sit in the shade, greet passersby, enjoy the view, and occasionally save somebody's life. But he told us his family didn't like it. They didn't like that he would invariably rush off to the river when called, even in the middle of the night.

"Are people upset when they realize you've kept them from killing themselves?" I asked.

There was a faint smile on his face. "Usually the women get very upset. But the family is grateful." He said there were a lot of students who tried. There was always a rush after exam results came out. Others were motivated by family disputes.

"Do people kill themselves because they can't marry who they want?" I asked.

He nodded. "Yes. There are plenty of love cases. It's mostly students and lovers."

He was staring at the barrage. I asked him whether he had ever lost anyone. He nodded without hesitation.

The defining moment of Sattar's lifeguarding career had come on a cold, foggy November morning, nearly fifteen years earlier. A crowded school bus had come across the barrage from the east, the driver speeding in the fog. In those days, Sattar said, there was no fence on the bridge. The driver had veered to avoid a pile of sand in the roadway, and the bus skidded out of control and crashed over the downriver side of the barrage. It was seven-fifteen in the morning.

"I dived in straight away," he said, pointing at a spot of water twenty feet from the bank. "Three boats charged, as well." The men dove and dove into the cold water, pulling kids to safety before going back to find more. Soon, they were finding only bodies.

"Now that I'm describing it to you, it's right there in front of me," Sattar said. "Everywhere we put our hands, we found them. Under the seats. I pulled out the body of one boy, and two others came with him." Out of 130 children on the bus, nearly 30 died.

The Wazirabad crash was a huge news story in Delhi, and Sattar received an award from the national government. There had been promises of money, too, but Sattar told us that had just been the chatter of politicians trying to look generous. They had never followed up.

But he didn't care. Lifeguarding was its own reward. He told us of one girl who had survived the crash. In a television interview, she had said it was thanks to Sattar that she was alive.

"I save lots of people," he said. "I've gotten used to it. But when that girl said that, it really touched me."

He shook his head, still deep in the memory. He had been shivering for a week, he said. The river had been very cold.

\\\\\\\\\

My original plan had been to find a canoe or a rowboat and run the Yamuna from Delhi to Agra, a journey usually made by bus. My waterborne arrival at the Taj Mahal—likely to a throng of local media—would open up an entirely new tourist route, and possibly lead to economic development along the water, and a renewed campaign to restore the Yamuna. *You're welcome.*

But my delusions faded fast. Just you try looking up *kayak* in the Delhi yellow pages. And although there are scores of whitewater rafting companies in the foothills of the Himalayas, I soon realized it was hopeless to try to entice them out of the mountains. I didn't have the money. Besides, they were *whitewater* rafters, not brown. Finally, there were all those dams on the Yamuna, and diversions, and dry sections. How do you raft a river that's not there?

On foot is how. I had learned there was a yatra under way. *Yatra* is a Sanskrit word for "procession" or "journey," and in this case meant a large protest march undertaken by a group of *sadhus*. Hindu holy men. They were walking a four-hundred-mile stretch of the Yamuna, from its confluence with the Ganga in Allahabad all the way up to Delhi, to demonstrate against the government's failure to clean up the river. If I could find the march, out there in the wilds of the state of Uttar Pradesh, I could tag along for a few days. What luck! Environmentalism, spirituality, a good hike—and it was free. Knowing I'd need some Hindi on my side, I asked Mansi if she wanted to come along. She agreed right away. She's a photographer, and photographers are always down for an adventure.

Before I left Delhi for the trip downstream, though, I went to see the source of the trouble.

The Najafgarh drain was once a natural stream, but even more than the Yamuna, it has been completely overwhelmed by its use as a sewage channel. With a discharge approaching five hundred million gallons a day, including nearly four hundred tons of suspended solids—yes, *those* solids—the single drain of the Najafgarh accounts for up to a third of all the pollution in the entire, 850-mile-long river. It is the Yamuna's ground zero.

We approached it on foot, picking our way around the hubbub of a construction site. There was a new highway bridge going up, bypassing the chokepoint of the road over the Wazirabad Barrage. Beyond the work area we found a footbridge that crossed the drain several hundred yards up from where it met the Yamuna.

The footbridge was a wide dirt path bordered by concrete parapets. Looking over the edge, we could see the wide, concrete-lined trough of the drain, perhaps two stories deep. A dark slurry surged along its bottom. The air nearly rang with the smell—that fermented, almost salty smell. Sewage. It was a smell somehow removed from actual feces. A smell that somehow distilled and concentrated whatever it is about feces that smells so bad.

I had smelled that smell before, but never had it smelled like it smelled that day at Najafgarh. It smelled so bad it gave me goose bumps. It smelled so bad it made my mouth water. The gag reflex scrambled up my throat, looking for purchase. I tried to take shallow breaths.

And yet.

I looked over the side again. Vegetation climbed the seams of concrete on the walls of the drain. Green, bullet-headed parrots flew over the dark water. Pigeons stepped and dipped on a concrete ledge. Butterflies flopped upward through the sunny air.

Moving to the downstream side of the bridge, I saw strings of flowers snagged on the electrical wires that crossed the drain. They had caught there when people had thrown them in. Even here, people offered.

And why not? Underneath the stink and the noise, the rationale unfolded. This was a tributary of the Yamuna. Are you not to venerate it, merely because it smells? Why not worship it, suspended solids and all? What could be more sacred than a river that springs from inside your neighbor's belly?

\\\\\\\\\\\

The temple of Maan Mandir stands on a craggy hill outside the small, tangled city of Barsana, seventy-five miles south of Delhi. They worship Krishna there, and you could do a lot worse. Krishna comes in the guises of an infant-god, a young prankster, a musician, an ideal lover, a fierce warrior, and—depending who you ask—an incarnation of the ultimate creator. With Krishna, you get it all.

Maan Mandir is the headquarters of Shri Ramesh Baba Ji Maharaj. Shri Ramesh Baba Ji—screw it, I'm just going to call him Shri Baba—was the guru who had launched the Yamuna yatra, and I had been granted permission to join the march on the condition that I visit him first. A reluctant guru-visitor, I had agreed only grudgingly. I was impatient to fall in with the yatra. Images danced in my mind of contemplative Hindu ascetics walking the banks of the Yamuna downstream from Delhi—the oxygen-starved, eutrophicated segment.

We had come to Braj, Krishna's holy land. Braj straddles the boundaries of several Indian states, at the middle of the so-called Golden Triangle formed by Delhi, Jaipur, and Agra, and is one two-hundredth the size of Texas. It was here, way back when, that Krishna spent his days herding cows, stealing butter, and having sex with milkmaids.

So it is hallowed ground, and when you consider that almost every hill and pond and copse of trees in Braj is paired with a story of one of Lord Krishna's frolics or flirtations, you begin to understand the environmentalist possibilities of Hindu belief. The very landscape of Braj is sometimes thought of as a physical expression of Krishna. And through it flows one of his lovers: the goddess Yamuna. In the temples of Braj, she is the holiest river of them all.

So the question isn't why Shri Baba had launched the Yamuna yatra, but why he hadn't done it sooner. Perhaps he was busy trying to protect the sacred hills and ponds of Braj. These were every bit as endangered as the Yamuna herself, and Shri Baba, in addition to pursuing a successful guruhood at Maan Mandir, had made local conservation into a specialty—restoring ponds, protecting forests, fighting illegal mining in the hills, and establishing retirement homes for cows. (Not so ridiculous if you think cows are sacred.)

The embodiment of deities and sacred history in the natural world would seem to give Hinduism a huge leg up on Christianity in the eco-spirituality sweepstakes. St. Francis notwithstanding, Christianity has tended toward the anthropocentric. Our holy figures are all human, and live in the human sphere, which—some people argue—explains the West's rapacious approach to its environment. Perhaps things would have been different if God had given Jesus the head of an elephant. And you know we Christians would have an easier time connecting to the rest of nature (and less trouble stomaching evolution) if there were a monkey in the Holy Trinity. Alas, we have no Ganesh, and no Hanuman.

Even worse, Christianity spent centuries promoting the idea that wilderness was either fodder for our dominion or a source of evil. The Devil was not in the details; he was in the woods. Of course, that's not true anymore. Now, we *love* the woods, love nature, and save our fear and abhorrence for the dirty and despoiled places, precisely because they no longer count as natural. I guess that pent-up Judeo-Christian negativity had to go somewhere.

So, for a long time we were semiotically handicapped in the West, and there was no chance of us worshipping our forests. (What are you, an animist?) Besides, the world of forests and rivers and mountains was not the world that counted. All that mattered was the world that came after this one, a Kingdom that needed no conservation.

But don't get all dewy-eyed about the alternatives. It seems humanity will find a way to ruin its environment, whether or not it's holy. The funny thing about vesting the physical world with divine meaning, as in Hinduism, is that the world can retain its sacred integrity whether or not it gets treated like crap.

Years earlier, in my visit to Kanpur, I had seen pilgrims taking bottles of Ganga water home with them to drink as a curative—a curative laced with sewage and heavy metals. When I asked one man about the quality of the water, he told me he wasn't worried. "It can't cause disease," he said. "Because Ganga is nectar. It can't be made impure."

And because a holy river has such purifying power, it is actually the perfect recipient for all your most impure waste—sewage, corpses, and so forth—which by mere contact with the water will be cleansed. So there is no paradox in the state of India's rivers after all. Their very holiness speeds their ruin.

\\\\\\\\\\\

From the crown of its ridge, Maan Mandir commands a blinding view of the surrounding plain. To the west is Rajasthan, hills rising against the horizon. Our media handler, a skinny sadhu called Brahmini, showed us around the temple and down to the lower buildings, where we would be staying that evening. His manner was gentle, almost shy, and although he spoke with a faint lisp, his English was good. He used it to provide a detailed and unceasing account of Shri Baba's work.

"Shri Ramesh Baba Ji Maharaj is the greatest saint of Braj," Brahmini said. "In fifty-eight years, he never leaves Braj. When he came, there were robbers at Maan Mandir. They gave troubles to Shri Ramesh Baba Ji Maharaj. They threatened him and brought twelve guns. But Shri Ramesh Baba Ji Maharaj didn't yield. He's doing so many good works for India, specifically

Braj. Braj has so many sacred places, but they are in a state of immense destruction."

I perked up when he got to the Yamuna. "Yamuna River is also in very bad condition," he intoned. "From New Delhi fresh water is not coming to Braj. It is stopped at the dam at Wazirabad. And instead of water, only stool and urine is coming to Braj. So yatra started two weeks ago in Allahabad, where Yamuna has confluence with Ganga. When yatra gets to New Delhi one month from now, millions of people will come to protest to the prime minister."

Stool and urine, I scribbled in my notebook. *Millions of people. Prime minister.*

"Shri Ramesh Baba Ji Maharaj's programs are not just for Braj," Brahmini said. "Not just for all of India. But for all of the world." He emphasized more than once that they accepted no money from the people who came to Maan Mandir, that free meals were given to all comers.

The most important part of their work, he said, was in the chanting of the holy names of God—specifically those of Krishna and of Radha, his lover and counterpart. Radha, milkmaid of milkmaids, was Krishna's true love when he roamed the hills of Braj—never mind that she was married—and their relationship was so important to these particular followers of Krishna that they rarely spoke of one without the other.

"So much power is in the holy name of God," Brahmini said. "You want to make sure that as many people hear the name of God as possible." Maan Mandir had been distributing megaphones to devotees in small villages, so they could circulate through town every morning, chanting Hare Krishna, spreading the names of God. The program had reached thirty thousand villages so far.

I took a moment to mourn a million quiet village mornings ruined by amplified chanting. But Brahmini assured me it was worth it. "People and animals are salvated only by hearing it," he said. "The entire atmosphere of the village is purified."

Holy names could do more than purify village life. They were critical for the broader environment, a spiritual action necessary to confront the irreversible destruction predicted by scientists. "Only by chanting of holy names, the future and environmental problems can be saved," Brahmini said. "He was a great environmentalist also, Lord Krishna was."

In the evening we went to see Shri Baba preach. The sermon—or maybe it was a concert—took place in a breezy, square room in one of the buildings down the hill from the temple. The crowd was entirely Indian; Maan Mandir didn't seem to be attracting any aging hippies or Silicon Valley dropouts. Shri Baba wandered in and sat on a low stage in front. He was in his late seventies but looked much younger. He had great skin. He was bald, with a perfect globe of skull that crowned an expressionless, hangdog face. He preached in Hindi, his voice low and strong, measuring his sermon with long pauses. As he talked, he noodled on an electric keyboard, and every now and then the music would take over, a drummer and a flutist would start up, and Shri Baba would shift seamlessly into song. His best move, which he pulled once or twice per song, was to let his melody soar into a high, long note: at this cue, the entire room would raise their arms and scream, an entire army of Gil Seriques. *AAAGGHH!*

\\\\\\\\\\

Early the next day, we went to see the morning sermon up at the temple itself. Brahmini and Mansi and I climbed the stairs through the trees to the top of the ridge, toward an impossibly brilliant sky. Outside the temple, Brahmini led us into a small garden, in the middle of which stood the statue of a blue-skinned woman. It was Yamuna herself, a faint smile on her face.

The temple was older and sparer than the buildings down the hill. It had a stone floor, cool under our shoeless feet, and unglazed windows looking out over the countryside. Mansi sat with the women, and Brahmini and I walked to a crumbling chamber adjoining the back of the room, where he could translate the sermon without disturbing everyone else. He had brought a handheld digital recorder, into which he would speak his translation. Later, he said, he would send the audio file to a devotee in Australia, who would transcribe it and post it on the Internet. They did this every day.

Shri Baba was sitting on another low stage facing the audience. He spoke. Brahmini leaned over to me so I could hear him as he murmured into the recorder.

"The greatest mental disease is attachment," he said. "Suppose a man is attached to a woman."

I sat up.

"Don't see the outside," Shri Baba told us. "See the inside. The body is full of bones, blood, urine, and stool. It gets old and dies." Brahmini's translation was rhythmic and precise. "There are nine holes in the body," he said. "Only dirt and pollution is coming out. *And think about that stool.*"

That was the key, according to Lord Krishna. "If you see the errors in the object, in the body," Shri Baba said, "your attachment will be destroyed."

I decided to give it a try. I thought about the Doctor, to whom I was still most abjectly attached. I thought about how she was full of stool and urine. About how she was nothing but flesh and bone. About how she would grow old and die. I saw her in a hospital bed, old and dying, full of stool and urine. A tourniquet of compassion seized me across the chest. My eyes filled with tears. It wasn't working.

Shri Baba was still talking. He wanted to get some things straight about stool. He was, dare I say, *attached* to the topic. There were twelve kinds of it, he said, and proceeded to lay out the whole taxonomy, stool by stool. The body was a factory of stools, he said. It was folly to perfume and beautify something so polluted.

I know he was just trying to help his sadhus control their libido. But seriously, why so down on stool? Is our human plumbing really so vile? And wasn't the Yamuna itself full of stool and urine?

I sat back, tuning out. As Shri Baba segued into a disquisition on lust, I watched two pigeons fornicate enthusiastically on a ledge above the doorway. A third pigeon arrived, and there was a fight, and then some more pigeon sex. It was hard to tell the sex from the fighting.

The sermon went on, in the gentle, alternating monotones of Shri Baba's words and Brahmini's translation. In a daze, I saw a fly circle out of the air and land on my forearm. I watched its head of eyes pivot back and forth. Then, hesitant, it lowered the mouth of its proboscis, and touched it to my skin.

\\\\\\\\\\

"Baba is calling you," Brahmini said, and we went in for our audience.

Shri Baba was sitting on a small dais in a long, bright chamber on the temple's upper floor, profoundly expressionless, profoundly bald, cross-legged. We put our hands together and sat at his feet. It was like the scene near the end of *Apocalypse Now*, when Martin Sheen meets Marlon Brando, except Shri Baba wasn't scary like Colonel Kurtz, and it was daytime, and I wasn't there to kill him. A dull roar of drumming and chanting emanated from downstairs.

He began talking in Hindi. I had feared he would tell us that only by the chanting of holy names could Yamuna be "salvated," but I detected a practical mind-set even before Brahmini started translating. Between my few words of Hindi and the language's liberal borrowing of English, I could get the gist. *Yamuna. Eighty percent. Water. Wazirabad. Twenty percent. Government not honest. No awareness.*

Brahmini translated, and then indicated that I should ask some questions.

I told Shri Baba that I understood the Yamuna was important because of its connection to Krishna. But what about places Krishna had nothing to do with? What about the rest of the world? Did Shri Baba care only for Braj?

"The importance of environment is all over the world," he said. "Without the non-human life there is no human life."

What Shri Baba really wanted to talk about was corruption. And he didn't mean it in the spiritual sense. He said India was corrupt from top to bottom, especially as related to the environment. The supreme court had decreed that fresh water should come to Braj through the Yamuna, and yet it didn't happen. The yatra's purpose was to confront that fact.

"Not even 1 percent of India's people think about purifying Ganga and Yamuna," he said. "People who make efforts for sacred works are crushed." He said a price had been put on his head during the fight to save the hills from mining. People had been kidnapped. Shri Baba had been poisoned.

He ran his hand over the dome of his head, his face still impassive. "But we don't fear death," he said. "I consider myself as dead."

\\\\\\\\\\\

We found the yatra that night, ten or fifteen miles southeast of Auraiya. They were camping in a grassy compound off a minor rural highway. The river was nowhere in sight. The roads and paths along its banks, I was told, had become almost impassable, especially for the support trucks. Sunil, the march's logistical manager, had chosen to take the yatra along Highway 2 for a little while. We'd get back to the Yamuna soon, he assured me.

It had taken us all day to get there. Mansi and I had traveled from Maan Mandir alongside a tall, dark sadhu with a grandly overgrown beard. He wore a plain white robe and his only possessions were a small digital camera and a nonfunctional cellphone. He had a kindly face, but we dubbed him Creepy Baba, for the way he kept trying to put his hand on Mansi's knee.

The idea had been for Creepy Baba to help us find the yatra, but over the course of multiple jeeps, buses, and one badly crowded jeep-bus, he proved blinkingly inadequate to the task. In Agra, he convinced us to board the wrong connecting bus, which we could only un-board after a quick shouting match with the driver and most of the passengers.

Oh god, said Mansi. Who knows where Creepy Baba is taking us.

Sunil picked us up in Auraiya and drove us to camp, where a pod of sadhus descended on us in greeting. Through Mansi, they asked me over and over how I had found out about the yatra. When I said I had read about it in a newspaper, online, they wanted to know *which* newspaper. I had no idea.

"Was it the *Times of India*?" asked one man.

I did know of the *Times of India*—and knew it was in English. "It could have been," I said.

"*Times of India!*" he cried to the assembled crowd.

Soon a cellphone was thrust into my hand. When, moments later, it was snatched away, I had been interviewed by a newspaper in Agra. I know this because Mansi later read me an extensive quote—attributed to me, but none of which I actually said—from a Hindi-language Agra daily.

The man who had asked me about the *Times* was called Jai. In Shri Baba's absence, he was lead sadhu on the march. Shri Baba never leaves the land of Krishna, and so would join the yatra only when it reached Braj. The sadhus were carrying a pair of his shoes on the march, though, so he could be there in spirit.

Jai had been following Shri Baba for ten years now. A former social worker, he lived at Maan Mandir and was an almost frantically amiable man. In Hindi, he apologized for not speaking English. In English, I apologized for not speaking Hindi. Not to be outdone, he made an elaborate panto-mime of seizing the air in front of my mouth, inserting it into his ear, and then raising his hands once more in apology.

No, I said. It is *I* who must apologize.

Conditions on the yatra were spartan but well managed. The tents were large, sturdy structures of green canvas, perhaps handed down by the British upon their departure in 1948. Each tent was strung with a single, blinding lightbulb hanging from an old wire connecting it to the generator. There was a steel water tank on a trailer, and a truck mounted with an oven for baking flatbread, and a crew of at least half a dozen guys whose job it was to drive ahead of the march, set up camp, and cook. All we had to do was walk.

There is a long tradition of political walking in India, and this particu-lar yatra happened to coincide with the anniversary of Ghandi's famous Salt March, the yatra of yatras. For more than three weeks in the spring of 1930, Gandhi and an ever-increasing army of followers marched toward the sea, where they would make salt from seawater, symbolically violating the Salt

Act imposed by Britain fifty years earlier. Along the way, Gandhi made evening speeches to the marchers and to the thousands of local people who
came to investigate.

Covered widely in the international media, the Salt March gave a huge
symbolic boost to the Indian independence movement, and put civil disobedience on the map as a major political strategy. The marches of the American civil rights movement were yatras. And it was in hope of a similar
runaway train of popular righteousness that Shri Baba and company had
launched the Yamuna yatra. So far, though, he had motivated somewhat
fewer marchers than Gandhi or Martin Luther King Jr. had. It was hard to
be sure in the dark, but I counted about twenty tents.

In the middle of camp, they were holding a *satsang*—a kind of group
discussion or teach-in. Two dozen people from nearby villages sat on the
ground in the garish light of a work lamp, while Jai talked over a microphone
connected to a pair of earsplitting loudspeakers.

"You are the owners of this country," Mansi translated. "Taxes are sup-
posed to perform for you, but they don't. You don't get what you deserve.
Come with us tomorrow morning. Come walk with us. Come with us to
Delhi."

\\\\\\\\\\\

At quarter past five in the morning, I became aware of the ground, and then
of the tent, and then of the sound of tiny cymbals clashing together. I
unzipped the collapsible mesh pod of mosquito netting—thoughtfully provided by Sunil—and stumbled out of my chrysalis into the dark of a new day.
Bats flickered overhead.

Jai was on the loudspeakers again. "FIVE MINUTES!" he said,
through a squeal of feedback. "IT'S OKAY TO CHANT GOD'S NAME,
SO LET'S DO IT!" He warmed us up with a piercing round of *Radhe
Krishna Radhe Sharma*. A couple of men in orange robes bumped around and
got in line behind the white pickup truck on which the loudspeakers were

mounted. Jai gave us our marching orders. "Don't get in front of the truck!" he said. There was some hollering, and they gave the truck a push. The driver popped the clutch, the engine burped to life, and just like that, the yatra was in business for another day.

There weren't more than twenty-five of us. We walked down the road, following the pickup truck, which was mounted with side-facing banners showing pictures of Shri Baba and the leader of the farmers' union, with whom Shri Baba had formed a strategic alliance. There were several union members among us, recognizable by their green caps.

We walked, passing misty fields of green wheat, and the day came up. I hung back a little, avoiding the sonic kill zone directly behind the truck, and settled into the rhythm of the march. Eventually Jai would tire of leading us in chants of *radhe-this* and *radhe-that*, and a combo of young sadhus would get out their drums and cymbals and improvise a vigorous set of Krishna-themed songs. Jumbled among them in the bed of the truck, a young man cradling a laptop with a data antenna and a webcam tried to throw together a live webcast. Once the musicians exhausted themselves, they would patch the speakers into the computer to play some pre-recorded Krishna hymns, and then some archival recordings of Shri Baba himself, his halting baritone resounding over the Indo-Gangetic Plain. Then we would pass through a village, and Jai would get excited again, and take up the mic, and the cycle would repeat.

At breakfast, eaten off leaf plates set on the ground by the side of the road, Sunil suggested that Mansi and I might prefer to ride in the pickup truck, or even in his jeep. It took some effort to convince him that we had come to the march with marching in mind.

The modest procession began again. A squat sadhu with a gray beard and a potbelly ranged to the side of the road, handing out handbills to onlookers, who gathered in small groups to read the news. Creepy Baba had his camera out. For every picture he took of the marchers or the country-side, though, he seemed to take two of Mansi.

Oh god, she said. He is so *creepy*.

Mansi wandered off to take some pictures of her own, and I found myself overtaking a trim man of sixty-some years, who was pushing a bicycle. He had been at the previous evening's teach-in.

"What is your country?" he asked, in cautious English.

"USA," I said, and he nodded and smiled. For his benefit, I decided to rock out my very best Hindi.

"Kya yatra acha hai?" *Is the yatra good?*

He nodded again. "The sleeping Indians must awake," he said, employing somewhat more English than I had expected. "Natural resources provide so many things to humanity, without which life cannot exist. The people in high power are interested only in a life of luxury. They must be dethroned."

His name was M.P. and he was a retired schoolteacher from a nearby village. His shirt pocket was weighted down with pens. He told me he was only joining the yatra for the day. I asked him if he thought the yatra would have any effect.

"If the task is great and the desire is good, it must have success," he said.

We walked a little farther.

"Do you believe in God?" he asked.

"No," I said.

He looked at me in smiling disbelief.

"But God gives air, water, so many things! To not respect him and believe in him is ingratitude."

I couldn't disagree. But I couldn't agree, either.

"I'm grateful," I said. "And I respect him. I just don't believe in him."

Our conversation was interrupted by Jai, who sprang from between us and bolted for the truck, jabbing the air with his fingers as he went. A new song had started, and he wanted to be in the mosh pit.

\\\\\\\\\\\

The more I thought about the Yamuna yatra, the more it blew my mind what a diverse range of traditions it interwove. There was the forceful nonviolence of Gandhi's political campaigns, of course. Then there was the ancient practice of religious pilgrimage, Hindu or otherwise. But since I'm an American, it was also impossible to spend any time with a troupe of scruffy, nature-worshipping activist holy men without stumbling, inevitably, over Henry David Thoreau.

It's hard to believe that a single, self-proclaimed slacker could be largely responsible for delivering us two of the best ideas of the last 150 years, but in Thoreau's case the slacker had some tricks up his sleeve. The first idea was that of civil disobedience, which Thoreau named and explained, and which he practiced in a limited, proof-of-concept kind of way. Half a century on, his ideas became a major inspiration for Gandhi, who credited Thoreau as an indispensable political strategist. (Another half century, and Thoreau's ideas found their way in front of Martin Luther King Jr.)

The second idea was that nature is good, and good for you. The best way for a person to strive for spiritual perfection, he argued, is through the direct experience of wild, untamed nature, which will free the mind from civilization's clotting noise. Thoreau wasn't the only one to espouse this idea—the 1800s saw a whole transcendental crew on the loose—but he expressed it with such humor and good nature, and in a way still so accessible to readers, that we might as well give him most of the credit. Every time someone goes for a run in the woods, or donates to the Sierra Club, or maxes out their credit card at REI, the man with the neck beard and the bean patch ought to get royalties.

If there was one way that Thoreau thought was best for getting in touch with the environment, it was walking. The guy made a yatra of every afternoon. He championed not only walking but also ambling, strolling, moseying, and above all, sauntering. In his essay *Walking*, he makes the wry assertion that "I have met with but one or two persons in the course of my life who understood the art of Walking, that is of taking walks—who had a genius, so to speak, for SAUNTERING." From those rhapsodic capitals, he moves directly to the task of blurring the line between loafing and sacred pilgrimage, arguing that to saunter effectively is to be on a holy journey to nowhere in particular.

The transcendental notion is that nature and wildness aren't mere symbols of cosmic truth, but its actual embodiment. So to steep yourself in them, it follows, is to allow your spirit to unfurl. But it requires more than your mere physical presence. You must saunter mentally as well, losing yourself in your senses, coaxing your mind to meander into nature as surely as your feet have. "What business have I in the woods," Thoreau asks, "if I am thinking of something out of the woods?"

But if you believe, as I do, that the concept of nature is pretty bankrupt these days, then the question becomes just where to meet your sauntering needs. It's easy to understand what's nice about a walk in the woods, but will less obvious places do the trick as well? Can you properly saunter across an oil sands mine? What about around a soy field? Is the tired ground of Spindletop somehow inherently unsaunterable?

Even Thoreau acknowledged that his own sauntering grounds—around Concord, Massachusetts—were only semi-wild at best, shot through as they were with logging trails, and old native American footpaths, and homesteads, and farms. And when he went to Maine, in 1846, in search of a truly primeval nature experience, Thoreau found himself badly freaked out by the more serious wildness he found. Nature wasn't always beautiful or sacred-seeming. It could be uncaring and inhuman. Nature could crush your spirit as surely as it could raise it. He was honest enough to admit it, though, and incorporated the experience into his ideas, deciding that the healthiest thing for a person was to have one foot in nature and one in civilization. Nature's American prophet preferred his wildness benign.

From our vantage point 150 years after his death, there are also darker undercurrents to be found in the environmental ecstasy of Thoreau's ideas. In *Walking*, he goes to great lengths to point out not only *that* he sauntered, and where, but also in which direction. He went West, and it was no accident. A deeply moral man, an energetic campaigner for the abolition of slavery, and a founder of civil disobedience, he was nevertheless a kind of imperialist. He believed in his civilization, and in its growth. "I must walk toward Oregon," he wrote from the East Coast. "And that way the nation is moving, and I may say that mankind progress from east to west." There was a continent to despoil and plunder, and in his good-natured, wildness-loving way, Henry David helped carry the flag.

Thoreau and company have something else to answer for, too, if you ask me. It has to do with that mystical experience of nature they were so keen on. On the one hand, they convinced the world that the forest was essentially good—an idea that sparked the environmental movement and continues to nourish it today. But there was a side effect. Because they also convinced the world that the way for people to benefit from nature's virtue was to *go get it*. Direct, individual experience was the ticket.

And so environmental rapture became yet another commodity to be extracted from the forest, or the savannah, or the ocean. And all the nature-loving, green-friendly people of the world are merely coveting the spiritual goods. We're desperate to preserve what we call nature, but maybe that's just because it's the best place we know of to go mining for enlightenment.

\\\\\\\\\\\

In the morning they walked, but in the afternoon the sadhus napped. You shouldn't overexert yourself in such heat.

We camped in a dusty grove a hundred yards off the road. After lunch, Mansi and I lounged in an open tent with Jai and Sunil and M.P., who had brought me the gift of a religious booklet called *Preparations for Higher Life*.

Sunil played his regular game: trying to get us to walk all the way to Delhi.

"You can't leave!" he cried. "We love having you here. We're going to put chains on you both!" He reached out and seized us each by an ankle.

Although he was a sadhu like everyone else, Sunil wore jeans and a shirt instead of robes. His parents hadn't liked the idea of him becoming a holy man, he told us. "At least dress normally," they had said, and so he did. The street clothes were appropriate to his air of easy competence. As yatra manager, he was the brains of the operation and by far the most sensible sadhu of the bunch. But he counterbalanced this with a maniacal sense of humor.

"Name change!" he shouted, pointing at me. "Gore Krishna!"

Mansi laughed. "He's calling you *white krishna*," she said. "He says you're substituting the pen and the camera for the flute."

Sunil rocked back and forth, slapping the floor of the tent as he laughed.

I asked them exactly what made a person a sadhu. Did you sign up? Did you have to be ordained?

"It's someone's way of life," Sunil said. "Someone who just wants to be with God, who wants to serve."

"Like you," said Jai. "You've come here. You're concerned for the world. Those who think for others are sadhus."

"So I'm a sadhu?" I asked. Could you become a sadhu involuntarily?

Jai ignored the question. "This is not an easy fight," he said. "Without pen and ink, it's not possible." And he wanted to make sure I had my story straight. "People used to drink Yamuna to purify themselves," he said. "Now you can't even touch it. Recently some pilgrims drank some Yamuna water and had to be hospitalized that same night." The villages along the river couldn't use it as a water source anymore.

"Can't government provide people clean water?" he demanded. "If the government can put a Metro train a hundred feet underground, it can do this." He chopped one hand against the other. Someone had to purify the purifier. "Until Yamuna is clean, we are not going to back off. This is higher than religion. Higher than human beings."

\\\\\\\\\\

Hiking with the sadhus is cheaper than taking the bus, and more scenic, but you will have to come to terms with crapping in the open, which for Westerners can be profoundly difficult. In the past, I had mocked people who worried too much about the bathroom arrangements of faraway places, but

I now saw that I was one of them. Worrying about bathroom access, I realized, was a fundamental expression of my cultural heritage. All of Western civilization, in fact, had been built on a set of technologies whose only purpose was to abstract the process of dealing with one's own feces. (Germany is the exception to this rule, with its lay-and-display toilet bowls.) In any case, I would happily have parted with a thick stack of rupees for some time alone with a chunk of porcelain.

Yatra-ing, you will also have to wrestle with the privacy issues inherent to certain parts of India: i.e., that there is none. There is someone hanging out, or working, or taking a nap, or a crap, behind every shrub and around every corner. I doubt this worries people who grew up in the Indian countryside; they don't mind that someone could catch sight of them squatting in a field. But for a white man from New York—and for an educated young woman from Delhi, Mansi confirmed—this is just *not okay*. So you need a system.

FIELD MANUAL FOR CRAPPING OUTDOORS WHILE HIKING WITH SADHUS

1. CHOOSE A TIME. Everyone else goes in the morning, but this may lead to co-crapping, or at least crap-camaraderie, among you and the sadhus, which you must avoid at all costs. Afternoon is best, when everyone else is taking a nap.

2. BRING YOUR OWN TOILET PAPER. Toilet paper does not exist for these guys, who instead take a small lunch pail of water along with them for the purpose of washing—a method for which you are not trained. So pack a roll or two. The drawback to toilet paper is that, since you will leave it behind, you are flagging your turd as your own. (You are, after all, one of only two people for miles around who believe in toilet paper.) Any sadhu who comes upon your work will therefore be able to scrutinize your method.

3. CHOOSE A LOCATION. You've got to work the sightlines. The second day on the yatra, for instance, I found a nice spot behind the ruins of a small, brick building that screened me off from the highway, as well as from a trio of truck drivers lounging by the dirt access road. That left forty-five degrees of exposure to the south, but with nobody in sight I liked my odds.

4. CRAP. Work quickly. This is no time for an e-mail check.

5. In standard North American al fresco procedure, this step would be **FLEE**. But I am introducing an additional, intermediate step: **PAUSE**. Pull up your pants, yes, but notice, as you do, how your turd, mere seconds into its existence, has already attracted several flies. Consider for a moment the miracle of this fact. That in the vast, hot, not particularly fly-infested flatness of the province of Uttar Pradesh, three or four flies will find your shit within in an instant and start laying eggs. That in the simple act of squatting behind a brick wall, you have provided untold wealth for a generation of minuscule beings, who will make your poop their home, getting born in it, burrowing through it, eating it, until one day, grown up, they will spread their translucent wings and leave your now desiccated turd behind, to search out new frontiers for their own children.

So, pause. You are walking with the holy men. Take a moment, and observe your humble pile of feces, and remember that in Delhi they worship entire canals of this stuff, and know that the wonders of the universe never cease.

6. FLEE.

\\\\\\\\\\\\

At dusk, the teach-in went mobile. We emerged from our naps, the musicians among us climbed into the pickup truck, and we set out en masse for the closest town.

It was a tiny village, modest to an extreme, a densely packed assemblage of brick and earthen houses. A buffalo or goat twitched on every other stoop. With its total absence of cars—and air-conditioners, and televisions, and electricity—the town must have represented the platonic ideal of small carbon footprint. But it was disorientingly poor. Not even a day's drive away, I had seen Delhi's cosmopolitan set sipping twelve-dollar cocktails in bars and lounges as chic as anything in Manhattan. Now we were here, on the other side of the planet, in a world fueled with patties of dried cow dung. The gulf—in culture, in economy, and above all, in class—was impossible to fathom.

I'll just say it. *India. Less tidy and homogenous than I'm used to.*

We hit town at full *Hare*, equal parts spiritual revival, political rally, and Mardi Gras parade. People came out onto their front steps to watch us churn down the narrow, muddy road. The sadhus chanted and sang and hollered for all they were worth, going all-in with every drum, loudspeaker, and cymbal they had.

"Come walk a few steps with us!" Jai blared over the PA. But he was upstaged by the farmers' union president, who had joined us that afternoon. I recognized him from his picture on the side of the truck, a glowering buffalo of a man with a slash of hair covering his mouth. At the edge of town, he climbed onto the truck and gave his best Huey Long impression, growling and yawping and waving his fist stiffly overhead. More water should be released into the river, he said. The sewage should be treated and diverted. It was a facile, rabble-rousing version of what I'd been told by boatmen in Delhi, by the coin collectors, by R. C. Trivedi. Everybody knows, in ways more or less sophisticated, how to restore the Yamuna: stop destroying it.

The music started up again, and the circus crawled out of town, trailing a crowd of fifty or sixty onlookers, all men. It quickly devolved into dancing and general hoopla, with a core group prancing around with epileptic fervor. The dancers included the union president's two bodyguards, each of whom was armed with one of the small-caliber rifles ubiquitous to Indian security guards. I did some complementary dancing of my own as the bodyguards jumped and gyrated, waving the barrels of their guns around with way too much abandon. And like this, we danced and chanted and cavorted our way out of town and back to camp.

We had not seen the river that day. Tomorrow, Sunil said.

I lay in the tent. I was rereading *Moby-Dick* . . . sort of. The Melville spell that Art had cast aboard the *Kaisei* had yet to wear off. In New York, I had borrowed the Doctor's old copy, a battered green paperback, and carried it with me ever since. Through Brazil, through China, on half a dozen twelve-hour flights. But I was still only two pages deep. It was hard to focus when I opened it. The text was overgrown with inky blue notes, written in the earnest script of an intelligent teenage girl. The Doctor had read it in high school. At nights on the yatra, lying in the tent, surrounded by the quiet clashing of cymbals, I thought of the curling spine of the book, of the paperclips lodged in its pages. I didn't even have it with me. It was in my luggage, stowed in the corner of a friend's house in Delhi. Some talismans you don't need to carry with you.

Instead, on my phone, I read the news from Fukushima. There had been an earthquake. And after the earthquake there came a tsunami. And after the tsunami came the meltdowns. Each time I looked, there was more news. Reactor cores that overheated. Reactor buildings that exploded. From a tent in the Indian countryside, I watched the evacuation zone blossom from two, to ten, to twenty kilometers.

A sickening familiarity hung over it all. I remembered Dennis, in the briefing room in Chernobyl, tapping his pointer on the image of the firemen's memorial. I saw his contamination map of the Exclusion Zone, a distorted starfish with a reactor at its heart. And now again. Another terrifying Eden erupting onto the landscape. Another fifty or hundred thousand people forced aside. Another ghost born to haunt the world.

\\\\\\\\\\

It was a noisy camp. The generator ran all night, and the sadhus, too, working in shifts to ensure no break in the cymbal tinkling and the *Hare Hare*–ing. Underlying it all was a low murmur of conversation that I eventually realized was a recording of Shri Baba giving a sermon. Mahesh, the young man with the laptop and the webcam, also had an MP3 player with external speakers. The first thing he did upon reaching camp every afternoon was to connect the speakers to the generator and start Shri Baba up. I came to find it almost comforting, this never-ending sermon, a low lullaby beckoning me into sleep against the hard, uneven ground.

Five in the morning again, and we woke up, Mansi and I each in our individual mesh pods of mosquito netting. For Mansi, the mosquito net served double duty as a sadhu net. We didn't put it past Creepy Baba or some other insufficiently detached holy man to come climbing in next to her, hoping to play Krishna to her Radha.

It was Mansi's last morning on the yatra. She had things to do back in Delhi. When she announced that she would be leaving, though, Creepy Baba had suddenly announced that he, too, needed to go to Delhi.

Oh god, Mansi said. I'll never get rid of him.

I emerged to the sight of the pre-dawn mortifications. There was always a sadhu balanced on his head in the tent across the way, or complicating his nostrils with yogic breathing, or inflicting himself with some other reverent difficulty. Somehow it always took me by surprise. When I leave a tent, I guess I'm expecting a campfire, or some beef jerky—not a holy man tied in a square knot.

More substantially, I wondered why there weren't any young environmental types kicking around. Where were the young green-niks of Delhi and Agra? R. C. Trivedi and Bharat Lal Seth had both suggested that secular environmentalists and Hindu spiritual groups were finally working together, after decades of pointless division. I had thought India was the place where someone was finally building the bridge between conservation and religion. And maybe so. But then where was everybody?

Mansi made her escape shortly after we started walking, hitching a ride to the bus station in Sunil's jeep. For a moment it looked like she would get away without Creepy Baba in tow, but at the last minute he realized what was going on. Running to the jeep, he piled in next to her, and they rode off together, Mansi staring at the ceiling.

A month later I would e-mail her from New York, asking how things had turned out when the yatra had reached Delhi. I hadn't found much in the papers.

She would tell me she had gone to see the protest. There had been nothing like the predicted half million people, she said, but there were sadhus from all over the world, and a strong showing from the farmers' union. Creepy Baba had said hello, and another sadhu had grabbed Mansi by the hand and dragged her up front to sit by the podium. There were speeches, and some loose talk about taking a sledgehammer to the dam at Wazirabad. But nobody in Delhi noticed.

"Sad," she wrote me. "There's so many of them, and zero press coverage." It seemed the media had exhausted itself earlier in the month, covering an anti-corruption hunger strike. In the end, Delhi would pay the Yamuna yatra about as much attention as it does the Yamuna itself.

\\\\\\\\\\

I continued with the yatra for a few more days. Because he spoke decent English, Mahesh installed himself as my new minder.

"I will be your translator!" he said, walking beside me, his arms swinging wide. "I am going to tell you SO many stories about Lord Krishna!"

An earnest, ever-smiling man in his mid-twenties, Mahesh looked more like a young computer science graduate than a sadhu, but his enthusiasm for Krishna was unrivaled. Thus was I treated to stories and digressions about Krishna and heaven, about Krishna and the boy stuck in the well, about how Krishna had been "naughty" and gone "thief-ing water." About how Krishna had ordered his minister to "make women more lusty," and had then vanquished the minister for criticizing him about it. About how Krishna had told the people to worship the forests and the hills instead of the lord Indra.

Mahesh on sin. Mahesh on how if you invoke Krishna you will prevent illness. Mahesh on sin, again. Mahesh on how he had so many sins. SO many sins! I began to wonder just what kind of sins we were talking about. The sin of attachment? The sin of being full of stool and urine? The sin of being member to a ruinous species? Or something else that shouldn't count as sin? Was his sin something he had done? Something he wanted to do? Something that had been done to him?

We walked. We sauntered. We made embalmed relics of our hearts. Mahesh on how with Krishna at your side, you will avoid car crashes at the last instant. How if someone tries to hit you, they will fail. How if they shoot at you, they will miss. So many things. SO many things, Gore Krishna! The stories of Lord Krishna are real history. This is not only scripture, no. It is scientific!

I began to wither in the grip of the sadhus' hospitality, guiltily dreading the second and third and fourth helpings of food, served with smiling insistence. My belly became bloated with lentils and bread. But I had no choice. When I chose to skip lunch one afternoon, it threw the yatra into a near uproar of concern.

And Mahesh's solicitude knew no bounds. Had I eaten? Had I eaten enough? Had I washed my hands? Had I used soap? Did I need a bath? Did I know I could take a bath under the spigot of the water tank? Would I like him to show me where this bath could happen? Was I *going* to take a bath? When was the *last* time I took a bath? I didn't like being reminded that out here I was less competent than a five-year-old.

"You have been to the forest?" he asked me after lunch.

"The forest?" I said.

"The forest! Did you not go, for letting? Toilet? Two or three days . . . "

"Oh. Yes." I gave my report. "I went yesterday and the day before. Don't worry. Three days without, that's not possible."

"Everything is possible!" he said.

And still we hadn't seen the river. Tomorrow, Sunil said. We'll get there tomorrow.

\\\\\\\\\\

At the same time, part of me became convinced of the sadhu life. The evening found a dozen of us crammed into a single tent, singing, drumming, clashing symbols. The young man leading the songs was the best singer and drummer on the yatra. He probably spent a good five hours a day in rapturous musical performance, whether on the pickup truck or in the evening, in camp. Tonight, he drew verses from an open book of scripture, knitting his brow as he strung out a melody, before throwing it out to the group, to repeat in a throaty, musical roar.

On my last night, sitting on the ground eating dinner, I was befriended by Ravi and Ramjeet, two fifteen-year-old sadhus who had worked up the nerve to try out their English on me. I wondered if they were runaways, but they said their families had both endorsed the move to Maan Mandir. They were inseparable. Like Gabe and Henry on the *Kaisei*, they had known each other since early childhood.

"Ramjeet is ideal friend," Ravi said, clapping him on the back.

Our conversation was soon joined by Ravinder—a hotel manager from Calcutta who spoke perfect English—and another sadhu, a fierce-looking man with a shaved head and goatee. After dinner, we retired to one of the tents to practice English and talk about how I should stay on with the yatra, go to live at Maan Mandir, and devote myself to Krishna.

I couldn't, I said. I had to go home. I was done traveling. I missed my friends. I missed my family.

"But God wants you to be here, wants you to be at Maan Mandir," Ravinder said.

Maybe I should have considered it. I'm sure there was a bedroll for me up in the temple building. I could sleep under a mosquito net in a row of sadhus. I could wake up to the words of Shri Baba, and a view over the hills and ponds of Braj. Was that so much less than I had to look forward to in New York? And I liked these guys. Usually I bristle at people trying to convert me to their religion, but sitting here I was somehow gratified by how they didn't insist.

In my eyes, they were also pioneers. They were among the few people in the world who would purposefully make a sacred pilgrimage to a river full of shit. They were expanding the sauntering possibilities of the human race. It was precisely because the Yamuna was so desecrated, in fact, that they were pursuing this additional reverence.

And because Shri Baba's strand of environmentalism doesn't require a sacred place to be pristine or free of human settlement, it lacks the kernel of misanthropy that nestles at the core of Western environmentalism. A paradox of the conservation movement is that it both depends on personal experience of nature for its motivation—and clings to the idea that modern humans have no place in a truly natural world. To include people in the equation—as with the loggers of the Ambé project—seems like a concession, or at best a necessary compromise. In the minds of many environmentalists, whether they admit it or not, the ideal environment would be one in which people were sparse, or absent. But the problems with this as a conceptual starting point are obvious. We're here. And Shri Baba and his sadhus, it seemed to me, offered the possibility of a different mindset, in which one could fight for the environment without pining for Eden.

Since I had Ravinder and company there, I tried to nail down a few Krishna basics. Could someone please tell me the exact words to the *Hare Krishna* chant?

"It is called the Harenam Mahamantra," Ravinder said, writing it out in my notebook in capital letters.

"Like we use soap for cleaning clothes," Fierce Baba said, "we use the Harenam Mahamantra to clean our minds. To clean ourselves from within."

We went from there, and soon the tent was in a holy tumult, with Ravinder and Fierce Baba debating and correcting each other's storytelling and theology, and Ravi and Ramjeet paying rapt attention, and piling more questions on top of my own. There were 330 million gods, I was told, with Krishna on top. He had created the others. But then a bunch of Krishna devotees *would* say that, wouldn't they?

They told me about Krishna. They told me about his life in Braj. They told me about Shiva turning into a woman so he could join the milkmaids and watch Krishna dance. And they told me about the love between Krishna and Radha, always about Krishna and Radha.

I asked them about attachment and self-denial. Why renounce worldly pleasure when Krishna had himself been such a playboy? This provoked an extended melee about whether Krishna had been a sadhu, and whether, perhaps by dint of successful sadhu-hood, through which he entered into godliness, he had earned a kind of free pass to enjoy himself as a young man in Braj. They were still debating when I left.

\\\\\\\\\\

Later that night, as all the sadhus slept, I crept out of my tent and walked to the nearby woods, for "letting," as Mahesh would call it. On my way back, I stopped in the patch of herdland behind the camp.

The full moon shone clear and cool and magnificently bright. It was a perigee moon—the closest, largest full moon in twenty years. The landscape shimmered in monochrome, the silent forms of cows and buffalo lying like dark boulders on the packed dirt. A cowherd rustled under a blanket.

The puzzle of Krishna and Radha flickered in my mind. I had found it hard to distinguish which of them the sadhus were actually worshipping, or if it was the relationship itself that commanded the deepest veneration, a love affair that was somehow a deity in its own right.

"Two bodies, but single body," Ravinder had said.

The love between Radha and Krishna had been no mere love. It was a love that had created the human love for God. It was the ideal connection between the human and the divine, embodied in the eternal romance of two young deities.

Eternal, but it didn't last. The time came when Krishna left the hills of his youth and went to fulfill his destiny as a warrior and lord. It is said that without Radha to animate his music, he laid down his legendary flute. Later, he married and had children with a princess in Dvaraka. I don't know what happened to the milkmaid Radha.

\\\\\\\\\\\

We walked. It was a good way to travel, watching the fields creep by, and smelling the air, and feeling the exhaust of passing trucks. There was still no Yamuna in sight—later today, Sunil told me—and we were hiking, as always, along the side of the highway. The trucks would blare their elaborate horns as they rushed past, sometimes melodious, sometimes earsplitting. It would be nice to think they were honking in solidarity with the yatra, but in India as in many countries, it is simply a part of driving to blast your horn when you are passing another vehicle, or being passed, or when you see something by the side of the road, or when you don't.

It was morning. I saw things. A dot of orange crossing an expanse of feathered grain. She turned, a woman, the tangerine cloth of her sari covering her head, just visible above the wheat. A sadhu with an ochre stripe painted across his forehead grabbed a handful of chickpeas from the edge of a field and handed me a sprig, and we ate the beans raw. The tall chimney of a brick factory, and another, and another. They drew dark plumes across the sky. We passed close to one. In a compound enclosed by walls of brick, men carted bricks to a kiln made of bricks under a tall chimney made of bricks. A peacock stood on a crumbling brick wall, iridescent in the dust. At the sound of our loudspeaker, the workers paused and watched us go, and we waved to each other.

"All the farmers, come to Delhi!" the sadhus chanted. "All the people, come to Delhi!" There were thirty of us.

A burst of parrots, and then a group of Sarus cranes coasted over our heads and landed in a field, each of them tall as a man, and more beautiful. Smooth, gray feathers lined their bodies, a flash of crimson around the head. In India, I hear, they are revered as symbols of marital happiness, of unconditional love and devotion. The species is classified as vulnerable, if not yet endangered.

The Doctor and I had been e-mailing. From New York to Linfen, and Delhi, and here on the road, sympathetic words echoed over the space between two diverging lives, building our goodbye.

"Please do not be sad," she wrote. "My love goes with you everywhere."

We walked.

I should be wrapping it up, I thought. The end of the story was somewhere nearby, just down the highway, where the road found the river. I should be ready for that moment. I should be thinking, reflecting on *my journeys in polluted places*, looking back across thousands of miles, distilling each location into its essence, saying what it all meant. Hadn't I already said it? That to chase after the beautiful and the pristine was to abandon most of the world? That the unnatural, too, was natural? Or was it the reverse?

It began on a train to Chernobyl. And I had tried to follow it, through oceans and mines and forests, past a chain of uncanny monuments to our kind. There was something I was trying to see. An asteroid was striking the planet. I just wanted to catch a glimpse. But it was impossible, because we *were* the asteroid. The world had already ended, with a whimper, and also it didn't end. Now we inhabit the ended, unending world that came afterward. The world with us. The world transformed. A crater yawns open from its center and a new nature floods across it.

It is the world as it is, not as we wish it would be.

But mostly, we walked. And I waited for that feeling. It found me in the mornings. On the road before sunup, the sadhus falling into rank, Potbellied Baba narrowly avoiding being run down by an oncoming truck, and we would set out. Someone had garlanded the pickup truck with a white flag, bordered in green—the fluttering standard of the farmers' union. I stayed in the back and watched us as we went, our tiny band of misfits, a ragged line of men, supposedly holy, straggling along the shoulder of the highway, down to Delhi, with the night's mist settling on the fields, and the sun just short of the horizon behind us, and it would find me. Somehow, that feeling. It started in the bones of my legs, and into my spine, and up the back of my neck, washing over my ears and face and my eyes, coursing through my scalp, streaming into the air above my head, lit with the fresh sun and then it was day. This happened. Every morning, this attack of gratitude, swarming over me, as we walked and walked, puppets to an uncertain music.

\\\\\\\\\

Only after we had been in camp for several minutes did I realize it. We hadn't made the river. I was leaving for Delhi in the morning. My Yamuna yatra had been completely Yamuna-less.

What the hell, Sunil?

"Gore Krishna!" he cried, and told me not to worry. We would see the river that afternoon. He had planned a field trip. I crammed into the jeep with half a dozen other people, and Sunil hit the gas.

As we headed west, the air became hotter, the earth tougher, the fields of wheat taller and blonder. Forty minutes and half a dozen quick stops for directions, and Sunil turned left down a small, barren gully. There were rowboats tied up in the dust. It was the edge of the floodplain.

We came out the bottom of the ravine and saw a stripe of water in the distance, beyond a wide sweep of sandy scrubland. The Yamuna at last.

But if I thought the sight of the river would be greeted with any reverence by the sadhus of the Yamuna yatra, I was mistaken. They seemed not to notice. Sunil was in the middle of a long set of stories that had reduced the car to uproarious laughter.

"What is he saying?" I asked Mahesh.

"He is telling a joke," he said, between gasps.

"Yes, I know," I said. "What's the joke about?"

"Yes!" he said, still laughing.

"No, Mahesh. *What was the joke?*"

"It is a . . . very different, something kind of joke."

Our destination was a temple overlook on the bluff opposite. We crossed a temporary bridge constructed of large steel pontoons and cracked timbers, and manned by a quintet of men sitting by a shack. The Yamuna glimmered in the late-afternoon light. On the other side, Sunil sent us shooting up the dirt road that climbed the hill, past low adobe houses, past a huge banyan tree, and finally parked by the temple. We spilled out of the jeep and walked by a pair of ruined towers to find the overlook. From the promontory, we could see green fields descending to the riverbank. A pair of fishermen plied the water in small boats.

This was Panchnada, the confluence that R. C. Trivedi had told me about. Nearly three hundred miles downstream from Delhi, four tributaries joined to feed the Yamuna a massive dose of new water, finally diluting the river's oxygen-starved flow. We could see the confluence in the distance—the confluences. From a confusing tangle of sinuous bends and meandering inflows, the Yamuna emerged clean at last—or cleanish—despite everything that had been done to it. It had been made to flow into the ground, to slosh along canals and up against barrages, to wind through the intestines of six-teen million people, to suffer any number of other transformations, and still it flowed. It may have to wait out humankind to find a less tortured course.

On the way back, about a hundred yards past the bridge, the deep, dry sand of the floodplain swallowed the wheels up to their axles. We got out and started pushing the jeep in different, uncoordinated directions. In the distance, we saw a truck having the same problem, and another jeep. The place was a car trap.

"Gore Krishna has caused us complications!" shouted Sunil, gunning the engine and spinning the wheels. (Don't look at me, Sunil—I wanted to walk.) Mahesh crouched by the tire, shoveling sand out with his hands. "With Krishna all things are possible!" he said. Behind every handful he scooped away, more sand ran in.

I wandered back to the pontoon bridge. The men sitting by the bridge-keeper's hut let me climb the ramp and stand on the steel plates of the roadway. I watched the river flow gently against the bridge, steel cables creaking with the strain. A fresh, sweet air came off the water. Downstream, a new bridge was under construction, a proper highway bridge, built on tall concrete pylons.

The men climbed the ramp to see what I was doing. Their English was almost as bad as my Hindi, but somehow we started a conversation. The bridgekeeper said his name was Tiwari, and he introduced me to everyone else. I took their picture and showed it to them.

Tiwari got it across that the bridge was seasonal. It was installed only for the dry months, from November to the middle of June. During the monsoonal flood, he became a boatman, ferrying people across on a square, flat-bottomed boat that he kept tied up next to the bridge. I didn't know how to ask him if he would still have a job when the new bridge opened.

They asked my name. Andrew, I said. *Andru*, they said. They didn't ask me why I was here, or who I was, or where I was going. They asked me if I had been on the river.

Not here, I said. I had been on the river in Delhi. I held my nose. They shook their heads and clucked their tongues in disapproval. But they were smiling. I shook Gorokhpur's hand—I think it was Gorokhpur—and his wizened face creased with laughter, and I laughed, too.

I realized that, among my five or six words of Hindi, I had several that might apply.

"Ye pani acha hai?" I offered. *This water is good?*

They nearly broke into applause. *Yes!* they said. *This water is good.*

"Delhi pani bahot acha nehi hai," I said, getting ambitious. *Delhi water is not very good.*

No, they said. *It's not.* One of them pointed upstream. *Panchnada*, he said, and his sentence dissolved in a filigree of Hindi. I pulled out my notebook and we started drawing. We drew the Yamuna, and the four rivers feeding it, the fingers of a watery hand, with the bracelet of a pontoon bridge riding up against its palm.

Once the five rivers come together, the water is good, they said. Tiwari gestured up and down the river, his arm outstretched. He had the English word.

"Purify," he said. "Purify Yamuna."

Upstream, the sun was setting. A temple on the rise of the opposite bank had descended into silhouette. The breeze off the water had cooled. I took a last look at the Yamuna. At the place where it became a river again.

Then I said goodbye to the bridgekeepers and started back across the floodplain, to where the jeep was still trapped in the sand.

ACKNOWLEDGMENTS

The people who appear in this book were incredibly generous with their time, their thoughts, and often their homes. I'm truly grateful to them, and to many others who remained behind the scenes.

For help navigating Kiev and Chernobyl, I must thank Olena Martynyuk, Damian Kolodiy, Dmytro Kolchynsky, and of course Nikolai and Dennis.

In Fort McMurray, in addition to Don and Amy, I was welcomed by Matty Flores and Corey Graham.

Port Arthur is a better town than it gets credit for, and I am especially grateful to Hilton Kelley, Steven Radley, Laura Childress, Peggy Simon, Charlie Tweedel, Duane Bennett, Rhonda Murgatroyd, Jeremy Hansen, Bryan Markland, and everyone at the Spindletop-Gladys City Boomtown museum. For background on the Gulf Coast and its refineries, I depended on conversations with Ilan Levin, Jim Blackburn, and Kristen Peek, among others. I especially want to thank Jane Dalton, Scott Dalton, Don Harlan, Kirk Boomer, and Walter Mattox for opening doors in the Southeast Texas oil industry. Adam Ellick of the *New York Times* not only nominated Port Arthur as a notable polluted place but also was generous and exacting with his reporting advice.

My visit to the Great Pacific Garbage Patch was only possible thanks to Project Kaisei's inclusive spirit, and I wish to thank Mary Crowley, Lenora

Carey, and the entire Ocean Voyages Institute. I may have questioned their approach, but I don't doubt their passion and commitment, and I am deeply grateful for my experiences as a deckhand on the *Kaisei*. It also seems important to thank any group of people whose company you still enjoy after three weeks at sea, which was true of the *Kaisei's* crew. Special thanks to Stephen Mann, who was largely responsible for our survival. I am also grateful to Nikolai Maximenko, of the University of Hawaii, and Bill Francis and Marieta Francis of the Algalita Marine Research Foundation. Gabriel Goldthwaite, Henry Whittaker, and Tim Jones very kindly reviewed the chapter, and Tim provided coordinates for the *Kaisei's* route.

In Brazil, I depended utterly on Gil Serique's limitless store of energy, enthusiasm, and knowledge. Anyone seeking a guide, translator, or drinking buddy in the Amazon should seek him out immediately. I am also indebted to Rick Paid, who was immensely generous with his time, as were Josenilson de Souza Guimaraes (aka Tang), Eric Einstein, Eric Jennings, Steven Alexander, Joe Jackson, Luiz Machado, Antonio Carneiro, Raimundo Carneiro, and everyone at the Ambé Project. Carolina Klauck Moraes provided invaluable logistical support from afar, as well as after-the-fact translation.

I'm very thankful to Cecily Huang for her research, translation, and logistics work in China. My deep appreciation also goes to the Han family, to Liu, and to the coal workers of a particular mine near Linfen. Thank you as well to Jonathan Watts, Andrew Jacobs, David Yang, Helen Couchman, Ami Li, Evan Osnos, and Ruth Morris.

Of the many people who helped me in India, I would especially like to thank Mansi Midha. For the warm welcome at Man Mandir and on the Yamuna yatra, I thank Shri Ramesh Baba Ji Maharaj and his followers, including Brahmini, Sunil, Jai, and Mahesh. Thanks also to Jason Burke, Kakoli Bhattacharya, Vimlendu Jha, Anand Bhaskar Rao, and Shruti Narayan.

In both the Pacific and Amazon chapters, I combined my research for this book with the production of television segments for the weekly newsmagazine *Dan Rather Reports*. I am most grateful to Mr. Rather for that opportunity, and for his words of encouragement; also to the management and staff at DRR, among them Wayne Nelson, Elliot Kirschner, Steve Tyler, and Andrew Glazer, who provided me with work, contacts, and advice.

As my agent, Michelle Tessler has been an ideal advocate for this project, and without her guidance and enthusiasm it would not exist. Colin Dickerman also believed in the book from a very early stage and made a home for it at Rodale. As for everyone else at Rodale, I can't thank them enough for their hard work and collegiality: Mike Zimmerman, Marie Crousillat, Brent Gallenberger, Aly Mostel, Maureen Klier, Amy King, and especially Gena Smith, without whose intelligence and judgment this book would be a confused mess.

It is impossible for me not to thank the New York Public Library and Wesleyan University for providing me with good places to write; as well as Keith Blount, who created Scrivener, the software in which this book was written. Professor Lesley Sharpe of the University of Exeter provided a clarified version of the epigraph. (The original version can be found in the subway station underneath Bryant Park, in New York City, which means I should thank the MTA as well.)

Many thanks to Paul Wapner for an engaging discussion of his work studying environmental politics, and to Jamie Tanner for drawing the beautiful maps that adorn the beginning of each chapter.

I'm truly humbled by the support I received from my family, especially Jane and Michael Blackwell, and from a deep roster of friends. They supported the dream of this book, indulged my stretches of writer's despair, and provided invaluable feedback on the manuscript. They include James Taft, Laura Driscoll, Matthew Blackwell, Alice Towey, Katie Ender, Scott Dalton, Lorena Sanches Agredo, Victoria Schlesinger, Sally Kim, Anamaria Aristizabal (who took me to Kanpur in the first place), James Higdon (who named the book), Naomi Goodman, Nick Bussey, Bryan Reichhardt, Anna and Ben Low, Brigid Rowan, Eric Laplante, Fleur Knowsley, Kate Pound, Alisa Roth, Hugh Eakin, Chad Poist, Jeff Cohen, Andrew Goldman, Kristen Cesiro, and the ever-vigilant Erin Lee Mock. As for Adam Bolt, I don't see how I could have managed without his help. I can't thank him enough for his collaboration and friendship.

Above all, I will never be able to repay James, Laura, Erin Lee, and Adam. Time and again, these four put a roof over my head, gave me a place at their tables, and showed me what friendship looks like.

INDEX